ROUTLEDGE LIBRARY EDITIONS:
ENERGY ECONOMICS

Volume 21

# THE BRITISH COAL INDUSTRY

# THE BRITISH COAL INDUSTRY

H. TOWNSHEND-ROSE

Routledge
Taylor & Francis Group
LONDON AND NEW YORK

First published in 1951 by George Allen and Unwin Ltd

This edition first published in 2018
by Routledge
2 Park Square, Milton Park, Abingdon, Oxon OX14 4RN

and by Routledge
711 Third Avenue, New York, NY 10017

*Routledge is an imprint of the Taylor & Francis Group, an informa business*



*British Library Cataloguing in Publication Data*
A catalogue record for this book is available from the British Library

ISBN: 978-1-138-10476-1 (Set)
ISBN: 978-1-315-14526-6 (Set) (ebk)
ISBN: 978-1-138-30356-0 (Volume 21) (hbk)
ISBN: 978-0-203-73092-8 (Volume 21) (ebk)

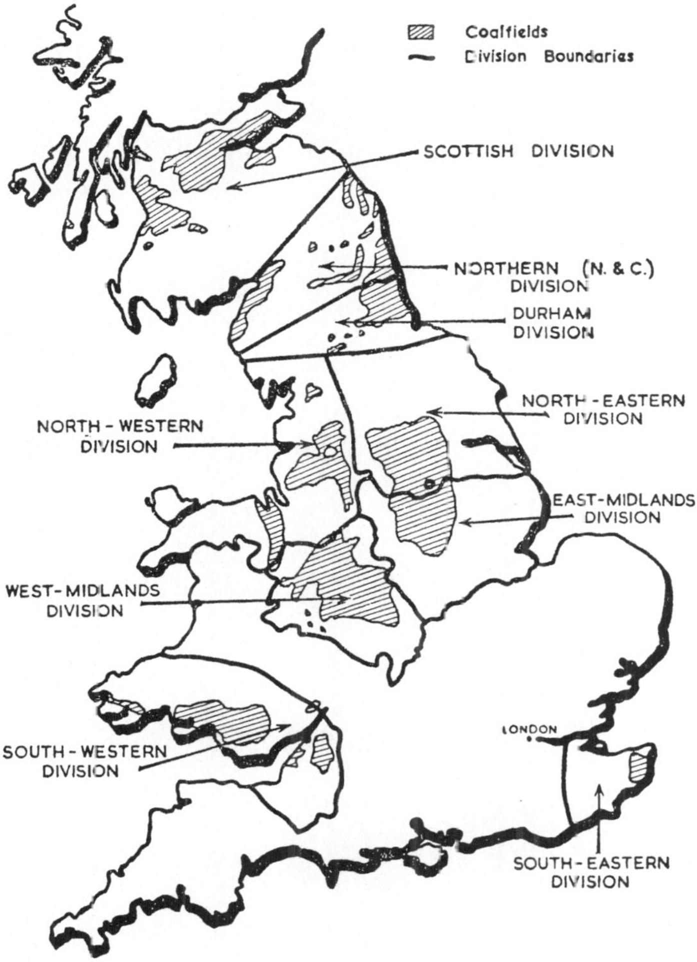


COALFIELDS AND N.C.B. DIVISIONS

# THE BRITISH COAL INDUSTRY

H. TOWNSHEND-ROSE

LONDON
GEORGE ALLEN AND UNWIN LTD
RUSKIN HOUSE MUSEUM STREET

FIRST PUBLISHED IN 1951



PRINTED IN GREAT BRITAIN
*in 12-point Fournier type by*
SIMSON SHAND LTD
*London and Hertford*

## Preface

In this book I endeavour to give a simple account of the way in which the British coal industry was nationalized, and of the new organization. My purpose has been to describe, not to criticize; for it would be presumptuous in me, as an employee of the National Coal Board, to attempt to evaluate what has been achieved. I hope that the book will be of use to new entrants to the Board's service, and of interest to all who are concerned with this great experiment in industrial organization.

The information contained in the book is based on material generally available at the beginning of 1950. I am indebted to the anonymous writers of the Annual Reports of the National Coal Board, not only for the wealth of information contained in those Reports, but also for the exemplary style in which the information is presented. I would also thank my colleagues for the help which they have given me in writing this account of the nationalized coal industry. I am grateful to the National Coal Board for the facilities afforded me in writing this book, but responsibility for statements of fact or opinion is mine, and what I have said does not represent in any way the official view of the Board.

The printer's house style has demanded the use of the letter 'z' in words, such as 'nationalization', where I would have preferred the use of the letter 's'. I have reluctantly agreed to this style.

H. TOWNSHEND-ROSE

*Chiswick, W.4*
*January* 31, 1950

## Contents

# Illustrations

Chapter 1

# BACKGROUND TO NATIONALIZATION

### *The Sankey Commission*

The desirability of nationalizing the British coal industry was the main feature of the reports emanating from the Sankey Commission, published in 1919. The Commission had been appointed in consequence of unrest in the industry after the 1914-18 war, and consisted of workers' and employers' representatives (not limited to the mining industry) in roughly equal numbers. A judge—Mr Justice Sankey—was chairman. The members of the Commission reached markedly divergent conclusions as to the future of the industry, and for that reason interest has attached to the report of the independent chairman.

The reports of the Commission were made in two stages. The first stage dealt with wages and hours of work and resulted in an increase in wages and a reduction in hours of work underground from eight to seven hours, plus 'one winding time' (see page 113). The second stage reports dealt with future organization of the industry.

The chairman in his report condemned private ownership of the minerals (i.e. the coal in the seams) because it had hindered development, wasted millions of tons of coal in barriers left between the properties of different owners, and resulted in the abandonment of coal and collieries because of lack of co-operation between different owners in drainage arrangements. He recommended that the unworked coal should be transferred to the State on payment of compensation. He also stated that relations between employers and workers in the industry were of such an unfortunate character that it seemed impossible to better them under the existing system of ownership of the mines. As he saw no hope of improvement he recommended State ownership of the mines on payment of compensation.

The workers' representatives were in substantial agreement with the chairman, but thought that coke and by-products plant associated with the collieries should also be acquired by the State. The mineworkers' representatives were not in favour of paying compensation to the mineral owners except in cases of hardship. The employers' representatives reported that they found no evidence that the coal industry could or would be as effectively conducted by the State as by private enterprise. They did not believe that nationalization would reduce the price of coal or prevent strikes, and they denied that there was any general antagonism between employers and workers in the industry. They considered that nationalization of the coal industry in any form would be detrimental to the development of the industry and to the economic life of the country. They agreed that the unworked coal should be transferred to the State on payment of compensation, and made a number of recommendations for improving the conditions of mineworkers, and for establishing machinery for conciliation and consultation.

On consideration of these reports the Government agreed in principle with nationalizing the minerals but decided against public ownership of the mines, and introduced the Bill which became the Mining Industry Act 1920. Part I of this Act established the Mines Department of the Board of Trade. Part II provided for the establishment of joint pit committees with power to make recommendations concerning safety, health and welfare, output, inspections and wages. There was to be a superstructure of district committees, area boards, and a national board, and the Board of Trade would have power to require compliance with recommendations made through this joint machinery. This Part of the Act never had effect. Part III of the Act contained a number of general provisions, one of which was a welfare levy on coal output.

### *Disputes and Settlements*

During the years 1920-26 the British coal industry was in serious economic difficulty and was troubled by disputes as to wages and hours of work. It is beyond the scope of this book to examine in detail the course of events associated with these disputes, but some knowledge of the terms of settlement reached is essential to an understanding of the extremely complex wage system of the coal industry.

During the 1914-18 war the Government had taken over full control of the mines and regulated profits and wages on a national basis. Immediately following the war the industry had prospered, largely because high prices were charged for coal exported; but competition from other coal producing countries, and reparations in the form of German coal, forced export prices down and the unsound financial state of the industry quickly became apparent.

In 1920, because of the rising cost of living, the miners demanded wage increases, and after abortive negotiations between the Mining Association of Great Britain representing the owners, and the National Federation of Mineworkers representing the miners, a strike of about three weeks resulted in a compromise settlement. The two sides of the industry then started negotiations with a view to establishing a new standard and providing a basis for regulating future changes. It was generally understood that, after Government control ended, the industry would not be able to afford the wage levels then prevailing.

The task was difficult because the wage system was already a complex one. The men were paid a basic wage, usually a pre-1914 standard, and to this were added various percentage additions as well as flat rate advances conceded during and after the war. The two sides were unable to agree as to what the new standard rate should be. They did reach some measure of agreement in principle that future changes in wage rates should depend upon a system of profit sharing, but were unable to agree as to the percentages to be assigned to wages and profits respectively.

In these negotiations the miners were particularly anxious to ensure that whatever settlement was reached should provide for the regulation of future wage variations on a uniform national basis. They wanted a system by which wage levels depended upon the financial position of the industry as a whole and not of individual colliery districts. The colliery owners were resolutely opposed to such an arrangement. They considered that it would be economically unsound and tend to keep unprofitable mines in operation when it would be better for them to close down.

The negotiations thus failed to reach agreement on three essential points: the new standard wage, the division of proceeds between wages and profits, and the question of national versus district wage regulation.

While the negotiations were proceeding the Government decided to end control on March 31, 1921. Negotiations then broke down and the owners gave notice terminating wage contracts on March 31; they later published the district wage rates which were to have effect from April 1. The new rates were unacceptable to the miners, and when the mines returned to private control on April 1, 1921, they stood idle. This national stoppage of the mines lasted three months, and the 'Terms of Settlement' with which the dispute ended became a permanent feature of the wages structure of the industry.

The Terms of Settlement fixed the new standard wage at the prevailing basis wage plus a uniform district percentage which was adjusted periodically in accordance with the proceeds of the industry as ascertained by districts. The district percentage was the result of a rather complicated calculation. First the aggregate proceeds of the sale of coal in the district during an agreed period were ascertained. From these proceeds were deducted (*a*) cost of wages at a 'notional' 1914 level, (*b*) 'notional' profits equal to 17 per cent of the 'notional' wages, and (*c*) costs of production other than wages. If any surplus remained, 83 per cent of it was added to the 'notional' wages, the total being expressed in terms of basis rates plus a percentage addition, and this was the new standard wage. It was provided that in no district should miners' wages be paid at lower rates than the 'notional' wages plus 20 per cent thereof. Allowances were also to be paid to ensure a subsistence wage to low-paid day-wage workers. It was also provided that if, in any period of wages ascertainment, the proceeds of the industry, after deducting (*a*) and (*c*) above, were insufficient to meet standard profits, the deficiency was to be carried forward as a first charge against any surplus in subsequent periods.

For the purpose of wages ascertainments the coalfields were divided into thirteen districts, and District Boards and a National Board, representative of both sides of the industry, were appointed with regulatory functions. The actual wages ascertainments were made by professional accountants representing both sides of the industry.

### *The Buckmaster and MacMillan Inquiries*

In 1923 the miners sought a revision of the 1921 agreement and a Court of Inquiry, chairman Lord Buckmaster, was appointed to report on the dispute which ensued. The Court reported that the wages of day-wage workers and of some, but not all, piece-workers, were substantially less than the equivalent of their pre-war earnings. In some circumstances the 1921 agreement resulted in owners' profits substantially in excess of pre-war profits while workmen's wages were less than the equivalent of their pre-war earnings.

After the submission of this report negotiations were resumed, and resulted in an agreement the principal effects of which were to modify the 1921 agreement in the following respects:

(*a*) The percentage of surplus proceeds allocated to wages was increased to about 87 per cent.

(*b*) The minimum percentage addition to the standard wage became 33⅓ per cent instead of 20 per cent.

(*c*) The scale of allowances necessary to ensure a subsistence wage was increased.

During the twelve months following the 1924 agreement the economic condition of the industry became steadily worse. Coal exports slumped and internal consumption declined as a result of trade depression. Wages and other costs of production increased and output declined. Many collieries operated at a loss, some worked short-time, and by July 1925 wages in every district were on the minimum percentage.

The Mining Association and the Mineworkers' Federation undertook a joint inquiry into the condition of the industry, but were unable to agree as to causes and the remedies to be pursued.

The Association believed that the main cause of the depressed condition of the industry was the high cost of coal production, and proposed a reversion to the eight-hour day and revision of the 1924 agreement. The Mineworkers' Federation were strongly opposed to these suggestions. A deadlock having been reached another Court of Inquiry was appointed, the chairman being the Rt Hon H. P. MacMillan, K.C. The Court expressed the miners' point of view succinctly: 'They represent that, unless they receive a remuneration for their labour which at best represents what they received before the war equated to the present cost of living, they cannot carry on the lives of themselves and their families, any more than the owners can carry on their collieries if they have to pay more than an economic price for labour.' The Court supported the continuance of a minimum wage and suggested that there was considerable room for improving the efficiency of the industry as a whole, particularly in the way of collective action on the part of collieries to enable facilities and resources to be used in common so as to promote economical working. The Court suggested that better results might be obtained from co-ordinating the industry with other enterprises; that improvement in the distribution of coal would result in lower prices; and that the discontinuance of the royalty system should be considered.

Faced with the prospect of another stoppage of work, the Government decided to institute a full inquiry into methods of improving the productive efficiency of the coal industry. In the meantime a subsidy was granted (£10 million was proposed but in fact £23 million was paid) to fill the gap between the level of wages fixed by the 1924 agreement and the lower level of wages that would result from the colliery owners' proposals.

*The Samuel Commission*

The promised inquiry was undertaken in 1925 by a Royal Commission of which Sir Herbert (now Lord) Samuel was chairman. The Commission had a double task: to suggest an early remedy for the serious economic state of the coal industry, and to make recommendations as to its future organization.

The Commission described the grievances which had exacerbated feelings between mineworkers and colliery owners. The miners maintained that the wage scale was at best no more than adequate, and for the lowest paid workers, and for all who suffered from short-time working, was plainly insufficient. In contrast, individual collieries paid large dividends and distributed bonus shares even in districts where wages were low. Considerable dissatisfaction was felt at the working of the wages ascertainment agreements. The miners believed that the proceeds of the industry were kept artificially low by the exclusion of coking and by-product plant and other activities, and that uneconomic prices were charged for coal transferred from collieries to associated businesses and sales agencies. The allocation of costs between capital and working accounts and between joint undertakings operated to the detriment of the miners, and the division of proceeds between profits and wages resulted in undesirable fluctuations in wage levels. With wages low, housing conditions bad, and the accident rate high, the miners protested that when the industry was in economic difficulty a further lowering of their standard of living was invariably regarded as a proper remedy. They complained of inefficiency in management, want of scientific methods of production, and excessive costs of distribution. They particularly objected to the payment of royalties in return for no service, and believed that the only way to improve the economic state of the industry was for the State to acquire and run it.

The owners considered that many of the grievances of the miners related to the past, and that others were deliberately exaggerated in order to support the political case for fundamental change. They denied that coal was transferred between associated businesses at unfair prices, or that capital was 'watered' by bonus shares. If wages were lower than they might be, or prices higher, it was due, the owners contended, to the men insisting upon hours of work so short that the working of the mines was unduly costly. They alleged that the men did not do their best to produce a satisfactory output even during the short hours they were at work. They complained of absenteeism, particularly among hewers, and an obstructive attitude towards mechanical methods and double-shift working. Disorganization and loss were occasioned by unauthorized stoppages for trifling grievances which could be disposed of by discussion. Increases in wages had more than kept pace with profits. The industry was running at a loss, and when the Government subsidy was withdrawn there was no alternative to longer hours of work and lower wages. The Institute of Mining Engineers strongly repudiated allegations of want of humanity and insufficiency of technical skill and scientific knowledge on the part of mine managers.

The Commission found that the industry as a whole had ceased to be remunerative and that, excluding the temporary Government subsidy, 73 per cent of the coal was produced at a loss. The Commission recommended that the subsidy be brought to an end and never repeated. As a solution to the immediate problem the Commission proposed that, if the existing working hours were to be retained, there should be a revision of the minimum percentage addition to standard rates of wages, which had been fixed in 1924 at a time of temporary prosperity. Only an immediate reduction of working costs in this way could save the industry from impending disaster. The wages of the lowest paid men were to be safeguarded by continuing the system of subsistence allowances, and the Commission left it to the miners to decide whether they preferred an extension of working hours with a less reduction in wages. The Commission considered it essential to have considerable variations in the rates of wages in various districts, but believed that national wage agreements should continue, and recommended that the two sides of the industry should meet together in order to arrive at a settlement.

In reporting on the structure of the coal industry the Commission emphasized the diversity in size of undertakings, geological conditions, geographical location, variations in working conditions, and competence of management. The Commission stressed the advantages to be gained by large-scale production, and considered that the amalgamation of many of the existing small units of production was both desirable and practicable, but might not be effected if left entirely to the action of the parties directly concerned. The Commission was opposed to nationalization of the mines, but recommended that the minerals should be acquired by the State on payment of compensation. Other matters with which the Commission dealt, were coal distribution and utilization, research, railway rates, recruitment, safety and health, housing, welfare, and the association of workers with management by means of joint committees.

After the report of the Samuel Commission had been presented the Mining Association and the Mineworkers' Federation held a number of joint meetings, but agreement was impossible. The owners sought a reduction in wages and longer working hours. The miners strenuously opposed any lengthening of the working day and refused to consider any reduction in wages in advance of reorganization of the industry. The Government failed to effect reconciliation and the most serious stoppage of work ever experienced in this country started on May 1, 1926, and lasted more than six months. The General Strike formed part of the events of those days.

During the course of the stoppage an Act of Parliament was passed permitting a temporary return to the eight-hour day (plus one winding time), and when work was finally resumed in November 1926 it was on the basis of district wage agreements, a reduced minimum wage, and longer working hours.

The bitter struggle of the year 1926 closed a chapter in the history of the coal mining industry: not because a satisfactory settlement of wage issues had been reached, but because disputes had been carried to the point of exhaustion. During the ten years following 1926 the country was beset with economic difficulties. The wages of the lowest paid workers in the coal industry were not adequate to maintain a reasonable standard of life, and unemployment and short-time working bore heavily upon those whose rates were above the minimum. In 1930 a 7½ hour day (plus winding time) was restored but it was not until 1936 that better economic conditions and the regulation of output enabled the miners to secure increases in wages and to win recognition of the principle of national discussion of wage issues.

After 1926 the history of the industry is concerned with the first tentative steps towards reorganization.

*Miners' Welfare*

Statutory provision for improving the conditions under which miners lived and worked was first made by the Mining Industry Act, 1920. This Act (among other things) imposed a temporary levy of a penny a ton on coal output, and the fund so created was applied to purposes connected with the social well-being, recreation and conditions of living of workers in or about coal-mines, and to mining education and research. The fund was administered by a Committee appointed by the Board of Trade, consisting of representatives of colliery owners and mineworkers, with some independent members.

The income of the Miners' Welfare Committee was augmented in 1927 by a levy of a shilling in the £1 on coal royalties. The proceeds of this levy were required to be used for the provision of pithead baths and facilities for drying clothes. An Act of 1934 reduced the output levy to a halfpenny a ton and extended until 1952 the period during which the levy was to be made. In 1939 the levy was restored to a penny a ton.

The 1939 Act also replaced the Committee which had administered the Miners' Welfare Fund by a Miners' Welfare Commission. The Commission was appointed by the Board of Trade and consisted of representatives of mineworkers, colliery owners and royalty owners, and some independent members.

The output and royalty levies, together with income on investments, yielded a total income which amounted at the end of 1946 to about £26 million. This money was applied by the Commission, and by the Committee which it replaced, in improvements in colliery areas which were of very great social value. The provision of pithead baths represented the outstanding achievement of the Commission, but in addition canteens were provided, miners' institutes, recreation grounds, first aid rooms, convalescent homes and rehabilitation and health services. Where the Commission did not themselves provide such services, they made grants which assisted other organizations to provide them. The Commission encouraged educational activities among mineworkers, and awarded scholarships and exhibitions to mineworkers and their children.

A further account is given at page 121 of the work of the Miners' Welfare Commission and of the way in which this work is being co-ordinated with the welfare activities of the National Coal Board.

*Working Facilities*

Although it was in the national interest that minerals, especially coal, should be worked, private ownership of land and minerals sometimes prevented such working from being undertaken. The owner of the minerals was able to impose an absolute veto on the working of them, and he or any other owner of land overlying the minerals could refuse to grant some essential working facility; or unreasonable terms, particularly as to the payment of royalties, might be attached to a grant of the necessary facilities. The only way in which these rights of ownership could be over-ridden was by the cumbersome and expensive procedure of a private Act of Parliament. In 1923, however, Parliament passed an Act—the Mines (Working Facilities and Support) Act, 1923—which was of general application. This Act empowered a Court—known as the Railway and Canal Commission—to grant orders facilitating (and in some cases restricting) the working of minerals. The Act applied to minerals generally, but in a later amended form it enabled anyone who was desirous of working coal to apply to the Commission for a 'working facilities order' where there was a danger of coal being left permanently unworked. The Commission had to be satisfied that it was expedient in the national interest for the order sought to be granted to the applicant. The facilities granted might include the right to let down the surface in the course of mining operations, the right to occupy surface land for various purposes connected with mining and associated activities (e.g. coal preparation, coke ovens, and brickworks) and various rights of way, water supply, drainage, etc. The right to work coal could also be freed from conditions in a mining lease which impeded efficient and economical working.

The Commission could also make orders restricting the working of coal, where it was in the national interest to do so, in order to ensure support for buildings and works on land which had no adequate right of support from underlying minerals. The Commission could, and usually did, order payment of compensation or some other consideration by the party who benefited from the order which the Commission made. Some public bodies, such as railway companies and authorities providing water and drainage services, had the benefit of special statutory provisions—usually referred to as 'mining codes'—for the protection of their works against the consequences of mining subsidence.

In 1949 the Railway and Canal Commission were abolished and their powers to grant working facilities and support orders were transferred to the High Court. The National Coal Board have to make application from time to time for working facilities orders, for although they own the coal in the seams they may need some mining facility (e.g. the right to withdraw support) affecting surface land which they do not own.

*Amalgamations*

In order to encourage the amalgamation of colliery undertakings the Mining Industry Act 1926 enabled colliery owners to prepare and submit to the Board of Trade schemes of amalgamation designed to secure greater economy and efficiency in working. The Act provided for total amalgamations, with complete fusion of colliery concerns, and for partial amalgamations providing for the joint exercise of powers and the conduct of business and operations. If the owners so wished the Board of Trade could assist in formulating these schemes. A minority of owners unwilling to take part in amalgamation schemes which met with general approval could be brought in compulsorily by 'absorption' schemes. Schemes approved by the Board of Trade had to be referred to the Railway and Canal Commission for consideration. If the Commission confirmed schemes after hearing objections to them, the schemes had statutory effect.

These statutory provisions for amalgamation had no substantial results. The only element of compulsion applied to minorities of coal owners who declined to join in schemes prepared by other colliery owners in the locality. A particular defect was that absorption schemes could not be made which dealt with marketing apart from production of coal.

The Coal Mines Act 1930 made a new approach to the problem of amalgamations by setting up a Coal Mines Reorganization Commission. The Commission's task was to further the reorganization of the industry by promoting the formulation of amalgamation and absorption schemes under the 1926 Act. If the owners did not respond to representations made to them, the Commission were empowered to prepare the necessary schemes for submission to the Board of Trade for approval and to the Railway and Canal Commission for confirmation. Before confirming schemes the Railway and Canal Commission had to be satisfied that they were in the national interest, were fair and equitable, would result in lowering the cost of producing and disposing of coal, and would not be financially injurious to undertakings proposed to be amalgamated. Statutory requirements in these general terms are very often difficult of proof on the facts of a particular case, and these provisions in the 1930 Act were no exception.

The Reorganization Commission tried by various means to stimulate the industry to formulate voluntary amalgamation schemes, but without success. They worked under a considerable handicap since it was generally considered that their compulsory powers would prove to be ineffective. The Commission prepared a number of experimental schemes, and contemplated a two-tier structure for the industry. Colliery undertakings would be amalgamated into units of optimum size by means of financial mergers ('total' amalgamations), and certain of the activities of these units would be controlled, in relation to wider areas, by loose forms of co-ordination ('partial' amalgamations). The only scheme which the Commission endeavoured to impose compulsorily was a partial amalgamation scheme affecting West Yorkshire. The Railway and Canal Commission declined to confirm the scheme, principally because it effected no amalgamation by fusion of companies and was therefore held not to be in accordance with the requirements of the 1926 Act.

The efforts of the Reorganization Commission met with determined opposition from the colliery owners. It was alleged that the widespread uncertainty to which the activities of the Commission gave rise adversely affected operations, and that valuable time and money, which could ill be spared, were involved in dealing with the Commission's proposals. The Commission themselves felt that the system of private ownership of minerals stood in the way of effective reorganization. In 1936 the Government asked the Commission to refrain from further action pending re-consideration of the whole matter.

The Reorganization Commission was abolished by the Coal Act 1938, and that Act effected the long delayed transfer of the coal to public ownership. On July 1, 1942, the coal became vested in the Coal Commission, and compensation amounting to £66.45 million was paid to the former owners. The Commission were not allowed to engage in mining operations, but they were required to exercise their functions, as 'landlords' of the coal, so as to promote the interests, efficiency and better organization of the coal industry.

The functions of the Reorganization Commission were transferred to the Coal Commission, but a new procedure was introduced for dealing with compulsory amalgamations. The Commission had first to satisfy the Board of Trade and Parliament (by provisional order procedure) that it was desirable in the interests of efficiency that a particular amalgamation should take place; they had then to satisfy the Railway and Canal Commission that a particular scheme was fair and conformed to the requirements of the 1930 Act. These new provisions scarcely had time to receive more than general consideration before they were suspended in consequence of the outbreak of war. Almost the last comment of the Coal Commission (in their 1945 report) was 'that there was no part of the British coalfields where further integration was not necessary and . . . in none of them was adequate further progress likely unless compulsory powers existed'.

*Central Selling*

Co-operative selling in the coal mining industry was considered by a Departmental Committee which reported in 1926. The

Committee contrasted the individualism of coal marketing in this country with the German system, in which cartels and syndicates closely regulated production, and sales were conducted by large selling pools instead of by individual collieries. The Committee reached the conclusion that the development of organized marketing was desirable in this country in order to avoid excessive competition as well as to effect economies and improvements in the marketing of coal and to help stabilize the industry. The Committee thought that lack of consolidation in the industry was a serious impediment to re-organization, and considered it necessary for amalgamations to consolidate the industry into a much smaller number of units.

The Committee recommended early development on a voluntary basis of local arrangements for organized marketing, and advocated the use of selling pools. The Committee envisaged a series of district marketing organizations as a later stage of development, with co-ordination of the activities of these district organizations as a final objective. So far as the export trade was concerned the Committee thought that stabilization of prices and output could only be effectively achieved by forming a marketing organization in each of the exporting areas and co-ordinating their activities by a central committee. A minority of members of the Committee thought that compulsion would eventually be necessary in order to establish organized marketing in the coal industry.

The Coal Mines Act, 1930, made temporary provision for a statutory system of production 'quotas' for individual mines in order to prevent over-production of coal, and for a statutory system of minimum prices in order to prevent cut-throat competition.

Colliery owners were constrained (compulsory powers being available) to prepare and submit for approval to the Board of Trade a central scheme for regulating the supply and sale of coal. By this scheme a Central Council representative of mine owners was established. This Council estimated the total national requirement of coal for a given period and divided the requirement into maximum production allocations for (originally) twenty-one coal districts. Executive Boards, representative of the owners, were established under District schemes, and each Board in effect divided up the district allocation among the collieries in its district. The Executive Board fixed a 'standard' tonnage by reference to recent output for each colliery, and from time to time fixed the colliery's 'quota' as a percentage of the standard tonnage. All the quotas added up to the district allocation. Penalties were imposed on collieries which exceeded their quotas, but quotas were transferable between collieries in the same district. The District Executive Board also fixed minimum prices below which various classes of coal could not be sold by collieries within the district. Interests of consumers were protected by national and district committees of investigation.

These provisions of the 1930 Act were severely criticized because they caused efficient collieries to work below capacity and kept in being uneconomic collieries—these being presented with a new and marketable asset in the form of their quota. This result was at variance with other provisions of the Act designed to promote amalgamations and the elimination of uneconomic collieries (see page 26). In point of fact the output and price regulation introduced by the 1930 Act was intended as a temporary measure to cushion the financial effect of reduced hours of work and to maintain prices and spread employment while amalgamations were being arranged or imposed. As it happened the amalgamation provisions of the 1930 Act produced no substantial results, but the temporary provisions for price regulation, more especially in the form in which they operated from 1936 onwards, resulted in the establishment of central selling schemes which were markedly beneficial to the industry and survived to the time of nationalization.

In its original form the system of regulating output and price contained a number of defects. The quota was a 'production', not a 'marketing' figure, and a colliery which produced its quota could not be sure of selling it; nor was there any certainty as to the price which would be received for the coal produced. District minimum prices (which were pithead prices) were either generally evaded—particularly by collieries unfavourably situated in relation to markets—or were set so low as to be meaningless. In addition, there was fierce sales competition between districts, particularly when the export market declined and 'exporting' districts started to undersell on the home market. In 1935 the scheme was amended so as to provide separate allocations and quotas for coal for inland and export markets; the Central Council was empowered to enforce price co-ordination among districts, and to grant separate district allocations for particular classes of coal.

The statutory selling schemes which eventually evolved stemmed from the arrangements for price regulation introduced by the 1930 Act, but it was one of the terms of settlement of the 1935 wages dispute that, in order to put the industry on a proper financial basis, the coal owners should establish in each coal mining district complete and effective control of the sale of coal. The Government stipulated that schemes should (*a*) cover all coal owners in each district and have a measure of permanency; (*b*) effectively prevent inter-colliery competition; and (*c*) be so drawn that evasions could not take place.

The new schemes became effective from August 1, 1936, three main types of scheme being established. In the first, all the coal produced in a district was sold not by the individual owners but by one Coal Selling Association acting for the whole district. In the second type individual coal owners continued to sell their coal, but sales were controlled by a Committee which regulated price, quantity and destination of the coal sold. The third type of scheme provided for central selling by groups, the activities of the group being co-ordinated by the Executive Board for the District. In 1938 a Central Appeal Tribunal was established to which appeal lay by anyone aggrieved by a decision of a Committee of Investigation regarding the administration of a central selling scheme.

The full effect of central selling schemes was obscured by abnormal war-time conditions, but they were undoubtedly of great value to the coal industry in the immediate pre-war period in securing an economic level of prices and so maintaining the profits and wages of the industry.

*War-time organization*

In the early part of the 1939-45 war large numbers of men left the industry to join the services or work in other industries. As a result coal output fell until production was inadequate to meet the needs of the expanding war economy. The Government took action by Essential Works Orders and other means (see page 103) to check the flow of manpower from the mining industry, and controls were introduced (see page 134) which limited allocations of coal to industrial and domestic consumers. In 1942 the Government decided 'that a system should be developed by which questions of wages and conditions of service in the coal mining industry would be dealt with on a national basis and by a properly constituted national body'. In consequence of this decision the National Conciliation Board for the coal mining industry was established, consisting of a Joint National Negotiating Committee representative of mine owners and mineworkers, with a National Reference Tribunal for deciding matters on which the Committee failed to reach agreement. War-time changes in the wages and conditions of service of mineworkers are described in chapter 6 of this book.

In order to ensure maximum production in support of the war effort the Government took control in 1942 of the operation of all coal mines. This control was exercised through a Controller-General, assisted by Directors concerned with production, labour matters, coal distribution, and finance (including coal prices). A war-time National Coal Board was established (not the Board of the nationalized industry) for the purpose of considering general planning of production, increased efficiency, the provision of supplies, the maintenance of manpower and labour productivity, and the safety, health and welfare of mineworkers. Regional Controllers and Regional Coal Boards were appointed for each coal producing area, and every colliery undertaking was required to nominate a single person to be responsible for receiving and carrying out directions given to him on behalf of the Controller-General. Pit Production Committees, which had already been established, were called upon to assist mine managers in securing maximum output.

Government control did not extend to the finances of the industry, but it was found necessary to introduce arrangements for stabilizing finances. At the start of the war the colliery owners gave an undertaking to the Government not to raise the general level of coal prices without Government approval. The Government in turn agreed that coal prices should be fixed at a level which would cover production costs and yield a reasonable profit; the Government emphasized that no subsidy would be payable in respect of coal production. It was, however, necessary to provide assistance for collieries which had lost European markets but whose continued output was essential in the national interest.

This assistance was given from a fund created by a general levy on colliery undertakings. The levy was limited to *6d.* a ton of coal supply and the resultant fund was administered by the Central Council of Colliery Owners and the District Executive Boards. Later this scheme was supplemented, and a levy was also introduced to meet the cost of the guaranteed wage.

In 1942 these arrangements were to a large extent replaced by the establishment of the Coal Charges Account. This Account was controlled by the Minister of Fuel and Power and the Treasury; its purpose was to finance part of the day-to-day costs of colliery operations and spread the burden of increased costs of production due to war conditions. All colliery owners were required to pay into the Account output levies which were increased from time to time during the course of the war, and amounted in 1945 to 15*s.* a ton. The Exchequer made advances to finance the Account. Colliery undertakings were allowed to draw from the Account the cost of wage increases (they were mostly flat rate increases per shift, and placed a greater burden on some collieries than others). Reduced output also increased costs in some Districts, and these Districts were granted a higher allowance out of the Account. War emergency assistance to necessitous collieries was also paid out of the Account. Notwithstanding a number of flat-rate increases in the price of coal (prices rose by 112 per cent during the war years), when the Coal Charges Account was closed at the end of 1946 it showed a deficit of approximately £25 million, and this had been met by Exchequer advances to the Fund.

*The Reid Report*

In 1944 the Minister of Fuel and Power appointed a committee of mining engineers, of which Mr (now Sir) Charles Reid was chairman, 'to examine the present technique of coal production from face to wagon, and to advise what technical changes are necessary in order to bring the industry to a state of full technical efficiency'. The committee reported the following year, saying that throughout the industry drastic technical reorganization was vitally necessary.

In comparing British mines with those of Holland, the Ruhr and Poland, the Report showed that during the period of about ten years ending 1936, the increase in coal output per manshift was 118 per cent for the Dutch mines, 81 per cent for the Ruhr mines, 54 per cent for the Polish mines, and only 14 per cent for the British mines. The report gave reasons why Britain had failed to keep pace with her Continental competitors. The British mining industry had been in a perpetual state of financial embarrassment and of uncertainty as to its future, and was unable to spend money on necessary major improvements, whereas the other European coal industries had been able to command adequate financial resources. In Britain private ownership of minerals had resulted in the construction of an excessive number of ill-planned mines of inadequate capacity, and the widely dispersed ownership of the collieries had militated against the closing of uneconomic mines and the concentration of mining operations. This was in contrast with conditions on the Continent, where State ownership of the minerals and highly concentrated systems of colliery ownership had resulted in more efficient exploitation of coal resources.

Most of the British colliery owners had been opposed to amalgamations and were wanting in vision. Mining engineers were discouraged by financial stringency, lacked technical independence, and many of them did not appreciate the extent of reorganization that was required. The mineworkers did not co-operate with their employers and, for fear of unemployment, had no enthusiasm for schemes of mechanization.

The Report then examined what was needed in the way of technical re-development, and recommended that the intensive Room and Pillar system of working (see Fig. 2, page 72) was the best means of obtaining high productivity, with Longwall Retreating as second choice. There was urgent need to develop types of mechanical coal cutters and loaders that were suited to conditions in British mines, and to promote knowledge and use of improved methods of roof control. Haulage practices in British mines were wasteful and called for revolutionary change; man riding facilities needed to be greatly extended. The Report showed that in the United States one haulage worker was employed for every 50 tons of saleable coal produced; in Holland one haulage worker for every 20-25 tons; while in Britain one haulage worker was employed for every 5 tons of coal produced. In the majority of British mines it would be necessary to undertake fundamental changes in lay-out, and to construct new underground roadways, before locomotive haulage could be adopted. There was considerable room for improvement in ventilation and lighting underground.

The increase in output which would result from the use of modern methods at the coalface and in the haulage roads would necessitate the replanning of loading stations and the maximum use of shaft capacity by means of improved winding arrangements. Central workshops serving groups of collieries were a necessity, and the general lay-out of surface plant needed planning from the point of view of amenity as well as of efficiency. Modernization of colliery power systems and the increased use of electricity would increase efficiency and result in the saving of fuel. Standardization was necessary in the sizing and classification of coal, and the methods of coal preparation in use were in urgent need of modernization.

The development of mechanical methods of mining made it essential to have better arrangements for the preliminary training of new entrants, as well as for general underground training and the specialized training necessary to produce the many skilled technicians required. The industry needed an educational system that would afford mineworkers opportunities for progressive advancement and attract men of ability from all classes of society. It was the duty of management to explain development plans and new methods of work so that mineworkers were encouraged to take a real and intelligent interest in their work, and better use needed to be made of machinery for the redress of grievances. Wages ought to be at least as high as in other industries demanding an equal degree of skill and effort. The mineworkers needed to have the assurance of secure employment and of knowing that proper care was exercised for their health and safety. In return the mineworkers had the duty of giving a full and fair day's work, of accepting proper discipline at the mine, and of abandoning lightning strikes. They should co-operate with the management in raising productivity and be prepared to abandon customs detrimental to efficient working.

The Reid Report stressed the fact that reorganization of the industry needed to be undertaken on the basis of coalfields and not of individual mines. Efficient working necessitated both long and short term plans, and this in turn required the employment of full-time planning staff. It was urgently necessary to sink new pits of adequate size, and with good roads suited to modern haulage systems. The Report recommended that all new and reconstructed mines should be equipped for double-shift working, and suggested a five-day week of eight hours a day for underground workers.

In conclusion the Report stated that it was evident that the necessary technical changes could not be satisfactorily carried through as the industry was then organized. The conflicting interests of individual colliery companies needed to be merged into compact units of manageable size, with full responsibility for developing the areas to which they related. It was essential to have, in addition, an authority with adequate powers for ensuring that the necessary mergers were made, for stimulating the preparation and execution of broad plans of reorganization, and for conserving the country's coal resources. Technical reorganization of the industry would take many years to complete, but steady advances in productivity per man and lower costs of production would be achieved each year. The final words of the Report are: 'But there is no time to be lost.'

* * * *

The General Election which was held in 1945 resulted in the formation of a Government pledged to nationalize the coal industry. A Bill for that purpose was formally read for the First Time in the House of Commons on December 19, 1945, and received the Royal Assent in the following year.

Chapter 2

# THE PROCESS OF NATIONALIZATION

*The National Coal Board*

The Coal Industry Nationalization* Act, 1946, established the National Coal Board, transferred the assets and various rights and liabilities of the industry to the Board, provided for the payment of compensation, and made various other provisions consequential upon the process of nationalization.

As originally constituted the National Coal Board consisted of a chairman and eight other members appointed by the Minister of Fuel and Power for a term not exceeding five years. The constitution was amended by the Coal Industry Act, 1949, which provided that the number of members of the Board, apart from the chairman, should be not less than eight nor more than eleven. Of eleven members at least three must be part-time, and more may be. The chairman and deputy chairman are appointed by the Minister—not elected by the Board members—and normally a Board meeting may not be held unless one of them is present. The 1949 Act empowered the Minister to appoint two deputy chairmen. Regulations have been made fixing the quorum of the Board at five and making general provision as to proceedings and meetings, but subject to these provisions the Board may regulate their own procedure.

In appointing members of the Board the Minister must choose from amongst persons appearing to him to be qualified as having had experience of and having shown capacity in the coal-mining industry, industrial, commercial or financial matters, applied science, administration, or the organization of workers.

The members of the Board, as constituted shortly after the amending Act of 1949 had effect, were as follows:

*Chairman*

Viscount Hyndley, G.B.E.

Formerly: Controller General, Ministry of Fuel and Power; Commercial

* As to spelling of 'nationalization', see Preface.

Manager of Mines Department; Director of the Bank of England; Managing Director, Powell Duffryn Ltd.

*Deputy Chairman*
Sir Arthur Street, G.C.B., K.B.E., C.M.G., C.I.E., M.C.
Formerly: Permanent Secretary Control Office for Germany and Austria; Permanent Under-Secretary of State for Air; Second Secretary, Ministry of Agriculture and Fisheries.

*Labour Relations Member*
Mr Ebby Edwards
Formerly: General Secretary, Mineworkers' Federation of Great Britain (now National Union of Mineworkers); President of the Federation; Member of the Royal Commission on Safety in Mines; Member of Parliament for Morpeth.

*Scientific Member*
Sir Charles Ellis, F.R.S.
Formerly: Scientific Adviser to the Army Council; Professor of Physics, King's College, London; Fellow and Lecturer, Trinity College, Cambridge.

*Finance Member*
Mr L. H. H. Lowe, F.C.A.
Formerly: Finance Director, Ministry of Fuel and Power and Ministry of Food; Partner in the firm of Thomson McLintock & Co., Chartered Accountants.

*Manpower and Welfare Member*
Sir Geoffrey Vickers, V.C.
Formerly: Director General, Economic Advisory Branch of the Foreign Office; Member of the Joint Intelligence Sub-Committee of the Chiefs of Staff Committee; Member of the London Passenger Transport Board; Partner in the firm of Slaughter & May, Solicitors.

*Board Member (full-time)*
Sir Eric Young, M.I.Min.E.
Formerly: Director of Production, Ministry of Fuel and Power; Managing Director, Bolsover Colliery Co.; General Manager, Barrow Collieries, Barnsley.

*Board Members (part-time)*
Mr J. H. Hambro
Managing Director, Hambros Bank Ltd.
Sir Geoffrey Heyworth
Chairman of Unilever Ltd.
Alderman Sydney Jones, J.P.
Chairman, Monmouthshire County Council; formerly an official of the South Wales National Union of Mineworkers.
Mr Gavin Martin
General Secretary, Confederation of Shipbuilding and Engineering Unions.
Sir Godfrey Mitchell
Chairman of Geo. Wimpey & Co. Ltd.

Board members are disqualified for membership of the House of Commons so long as they hold office, and they may not, without the Minister's approval, engage in trade or business or be company directors or trade union officials. Full-time members are normally appointed for a term of five years, and part-time members for a term of three years. The Minister may remove a Board member because he is unfit to hold office, or is incapable of performing the duties of his office, or because he has been absent from Board meetings (unless for some approved reason) for a period of more than six consecutive months. Members of the Board must disclose to the Minister any interest or securities which they hold in any undertaking carrying on a business similar to the Board's. They must also disclose at a meeting of the Board any direct or indirect interest which they have in contracts proposed to be made by the Board, and they may not take part in the deliberation or decisions with respect to such contracts.

The salaries and allowances paid to the members are determined by the Minister with the approval of the Treasury, but they are paid out of the Board's revenues. Pensions or gratuities may be paid to them or to their dependants on retirement or death. The salaries actually paid are £8,500 per annum to the chairman and £5,000 per annum each to the other full-time members. In addition an expense allowance of £10,000 per annum for the whole Board is provided. Part-time members receive a fee of £500 per annum.

The Board is an independent corporation and not an agent of the Crown or a Government department. Employees of the Board are not civil servants; their salaries are paid out of the revenues of the industry and not out of Treasury funds. Documents of the Board are public records, but this provision refers only to their preservation and safe custody, and does not give the general public any right of access to them. There is, however, a general right to inspect plans of workings of coal on payment of a small fee.

*Functions of the National Coal Board*

Responsibility for operating the coal industry is imposed by statute on the Board. Although the Minister has power to give the Board general directions, and Parliament is entitled to reports of their activities, responsibility for success or failure of the nation's key industry rests quite clearly upon the small group of men who constitute the Board. This statutory responsibility has inevitably had a powerful influence upon the nature of the organization which the Board have created.

The Board's duties as laid down by Parliament are (*a*) to work and get the coal in Great Britain; (*b*) to secure efficient development of the coal-mining industry; and (*c*) to supply coal of quality, size and price, and in quantities, as may seem to them best calculated to further public interest in all respects, and without showing undue or unreasonable preference or advantage to anyone. The Board may carry on any activities which they consider to be requisite, advantageous or convenient in connection with the discharge of their duties, or for the purpose of making the best use of the assets vested in them. They have the exclusive right to search and bore for coal in Great Britain, and they may produce, supply and sell coal and the various products of coal, e.g. coke, manufactured fuels, gas, and the vast range of coal by-products. They may also provide the goods and services which they need in the course of their various activities, or which they can advantageously provide because of their activities. For example, the Board may generate electricity for operating a colliery and sell any surplus electricity which they so produce. They may undertake training and research projects for advancing the skill of persons who are or may be their employees, or the efficiency of their equipment and methods, and the provision by others (e.g. universities) of facilities for training, education and research.

In discharging their duties the Board must direct their policy to securing the safety, health and welfare of their employees, and the benefit of their practical knowledge and experience. The Board must also do their best to ensure that they pay their way over an average of good and bad years. The Act of 1946 provided that the activities of the Board should be limited to Great Britain, except that they might sell or supply for export, but the Coal Industry

Act, 1949, empowered them to carry on any activities outside Great Britain permitted by Orders made by the Minister of Fuel and Power. Any such Order must be made in the form of a statutory instrument laid before Parliament, but Parliament cannot amend or annul the Order. The first Order made authorized the Board to carry on activities (except searching and boring for or working coal) in Northern Ireland, the Isle of Man and the Channel Islands; to sell coal and coal products on c.i.f. terms to buyers in the Republic of Ireland; and to sell by-products on c.i.f. terms, hold patents, etc., advertise, and maintain sales or purchasing agents, anywhere in the world.

The National Coal Board have not in fact an exclusive monopoly to work the coal in Great Britain. There is a limited exception by which anyone who undertakes activities other than colliery activities (for example, working a bed of fire-clay) is permitted to dig and carry away any coal that it is necessary to remove in connection with those other activities (for example, removing a 'roof' of coal in order to get at the fire-clay). The Board may license the working of coal present among other minerals but of so small value that the working of it is not likely to be undertaken except as ancillary to the working of those minerals. In addition the Board may license small mines in which the number of underground workers does not greatly exceed 30. At the end of 1949 there were about 400 of these licensed mines. Open-cast mining is not undertaken by the Board but by the Minister of Fuel and Power (through contractors) in accordance with powers conferred by the Defence Regulations (see page 86).

### *Ministerial and Parliamentary Control*

The Ministry of Fuel and Power was established by an Act of Parliament of 1945, and the Minister is charged with the general duty of securing effective and co-ordinated development of coal and other sources of fuel in Great Britain. The Coal Industry Nationalization Act and the amending Act of 1949 add to his powers and duties. His power of appointing and dismissing members of the National Coal Board is of obvious importance, though dismissal may not be resorted to arbitrarily. After consultation with the Board he may give them directions of a general character as to the exercise and performance of their functions in relation to matters affecting the national interest. In framing programmes of reorganization or development involving substantial outlay on capital account, and in the exercise of their functions as to training, education and research, the Board must act on lines settled from time to time with the Minister's approval. A formal Order by the Minister is necessary before the Board can engage in overseas activities. Parliament has imposed upon the Minister the duty of securing the prosecution of research into methods of advancing the safety and promoting the health of persons employed in or about coal mines.

The Board are required to afford the Minister facilities for obtaining information with respect to their property and activities, and they must furnish him with various returns and other information. They must make an annual report to him on the exercise of their functions and on their policy and programmes, and the report must contain any directions given to them by the Minister during the year, unless it is against the public interest to give publicity thereto. A copy of the annual report must be laid before each House of Parliament, and may be the subject of debate. The Minister also exercises a considerable measure of financial control over the activities of the Board as shown in Chapter 8.

Subsequent to nationalization of the coal industry, proposals for nationalization of other key industries necessitated provision for co-ordinating action where there was the likelihood of conflict or overlap. Thus, the Gas Act, 1948, requires the National Coal Board to consult with the Gas Council, as well as with any Area Gas Boards concerned, in planning and carrying out programmes of capital development and reorganization of activities relating to carbonization (i.e. the manufacture of coke and related products). In addition, schemes must be submitted for the Minister's approval providing for co-ordination in the national interest of the activities of the two industries in relation to carbonization. This obligation to consult on the subject of carbonization will exend to consultation with the Iron and Steel

Corporation when the Iron and Steel Act, 1949 (which nationalizes the iron and steel industry) has effect.

Another aspect of Parliamentary supervision of the National Coal Board arises in connection with questions in Parliament. During the war the Government exercised direct control over a number of industries, and Ministers answered detailed questions on matters concerning their day-to-day administration. This war-time practice, together with some concern as to the working of a nationalized industry, resulted in Members of Parliament seeking to ask very detailed questions as to the activities of the Board. It therefore became necessary for the Government to review the principles which were to govern questions and answers in Parliament concerning the work of the socialized industries. It was stated on behalf of the Government in December 1947 that where boards had been set up to run socialized industries on business lines a large degree of independence in matters of current administration was vital to their efficiency as commercial undertakings. A Minister would be responsible to Parliament for action taken in relation to the board of a socialized industry, or for action coming within his statutory powers which he had omitted to take; but he would not, in replies to questions in Parliament, or in letters, give information about day-to-day matters, because undue interference by Ministers would impair the commercial freedom of action of the boards.

### *Consumers' Councils*

The interests of consumers are provided for by the establishment of two Consumers' Councils, known as the Industrial Coal Consumers' Council and the Domestic Coal Consumers' Council. The members of these Councils are appointed by the Minister, and in appointing them he must consult bodies representative of the interests concerned. The persons appointed to the Industrial Council must be representative of consumers of coal, coke and manufactured fuel for industrial purposes or other purposes involving bulk supply, and of persons engaged in the sale or export of those commodities. The members of the Domestic Council must represent the interests of domestic consumers of coal, coke and manufactured fuel, and of suppliers of fuel so consumed. The persons appointed to these Consumers' Councils must have adequate knowledge of the requirements of the interests they represent, and be qualified to exercise a wide and impartial judgment on the matters to be dealt with. The Minister must also appoint representatives of the Board to serve on the Councils. The chairman of the Industrial Council is Sir Ernest Smith, formerly Director of a large engineering company and war-time Director-General of Gas Supply. The Council includes coal traders and exporters, iron and steel manufacturers, electrical and mechanical engineers, representatives of railway, gas and electricity authorities and co-operative societies, scientists and trade unionists. The chairman of the Domestic Council is Dame Vera Laughton Mathews, formerly Director of the W.R.N.S. Among the other members are representatives of local authorities, trade unions, co-operative societies, women's organizations, coal merchants, gas and electricity authorities, and a scientist. The Deputy Chairman of the National Coal Board and the Board's Director-General of Marketing are members of both Councils.

The functions of the Consumers' Councils are to consider, within their respective provinces, any matter affecting the sale or supply of coal, coke or manufactured fuel. Their deliberations may be of their own initiative or in consequence of representations made to them by consumers, and they must notify their conclusions to the Minister where they think action is necessary. They may also examine and report to the Minister on any matter which he refers to them. The Minister will consult the Board on representations made by the Councils, and if a defect is disclosed in the Board's general arrangements for the production, sale or supply of coal, coke or manufactured fuel, he may give the Board directions for remedying the defect to which they must give effect. Each of the Councils makes an annual report to the Minister and these reports are laid before Parliament.

The Consumers' Councils have a membership link with the Board, but in other respects they are quite independent of them. Their staff, offices and expenses are paid out of Treasury funds and not out of the revenues of the coal industry. Regional Councils may be appointed by the Minister to represent industrial or domestic consumer interests. In that case regulations will establish the relationship between the regional and national Councils and decide whether the reports of the regional Councils will be made to the corresponding national Council or direct to the Minister. No regional Councils have yet been appointed.

*Transferring the Industry*

The task of transferring the coal industry from private to public ownership was an extremely complicated one. Much more so than the transfer of the railway, electricity and gas undertakings, for the operational units of those industries were to a considerable extent self-contained and grouped into a few large units. In contrast the 1,400 odd collieries managed by more than 800 companies and individuals were in many cases operated in close association with other activities, particularly the manufacture of iron and steel. It was often very difficult to decide whether particular lands or buildings, or even items of office furniture, were used in connection with coal industry activities to an extent that justified transference in the process of nationalization.

The greater part of the coal industry was transferred to the National Coal Board on January 1, 1947 (known as the 'primary vesting date') but the complexity of the process of nationalization was such that many assets were transferred at later dates. Even by the end of 1949 the Board did not know everything that had been transferred to them, though they found themselves in possession of such unexpected assets as a milk round, a holiday camp and a cycle track. The word 'vesting' which is in constant use in connection with the process of nationalization is a legal term signifying the transference of ownership of property from one person to another. Hence, the 'primary vesting date' is the basic date on which ownership of the principal assets of the coal industry passed from the former owners to the National Coal Board.

The minerals—that is to say the coal in the seams as distinct from colliery equipment—had already been transferred to the Coal Commission by the Act of 1938, and it was a simple matter to transfer the ownership of this coal and of other property of the

Commission to the National Coal Board. This transfer took effect on January 1, 1947, and the Coal Commission ceased to exist soon after.

For the purposes of nationalization the assets of the coal industry were divided into three main groups. In the first group were the collieries themselves with their associated coke ovens, manufactured fuel plant, lands, stocks of products, electricity plants, railways, canal wharves, merchanting property, welfare institutes, and general equipment: all things which indisputably made up the main physical assets of the coal industry. These assets were transferred to the Board automatically without either the Board or the owners having to take steps in order to effect the transfer. In the second group were colliery stores, waterworks, wharves (other than canal wharves), colliery houses and farms. These were transferred if the Board exercised an option to have them or if the owners exercised an option for the Board to take them. Where such an option was exercised, whether by the owners or by the Board, ownership of the property was transferred without either party having the right to object to the transfer. The third group consisted of a miscellaneous range of property associated with colliery activities (brickworks are an example) which the Board might opt to have or the owners might opt to transfer, but in respect of which reference to arbitration was available if either party objected to such transfer. In relation to all these groups it was specifically provided that iron and steel works, and property used for supplying coke to such works, should not pass to the Board. A large number of railway wagons for carrying coal were transferred to the Board on January 1, 1947, and a year later they were transferred from the Board to the British Transport Commission when the railways were nationalized.

In order to ascertain which assets would be transferred automatically and which might be the subject of options, colliery owners were required to produce 'statements of interests' showing the various assets capable of vesting and any rights, such as easements and contractual rights, associated with them. They were also required to produce inventories of stocks and stores. In addition to providing for physical property to vest in the Board, the process of nationalization necessitated the transfer of intangible things, such as rights in patents and designs, and rights and obligations under contracts which were to remain in force. Particulars of assets and associated rights actually transferred to the Board are recorded in 'compensation units', and these are the basis of distribution of compensation in the manner described on page 51.

Parliament was aware of the need to prevent attempts to dispose of property so as to avoid its transfer on nationalization, and to ensure that property was properly cared for pending transfer to the Board, and appropriate provision to this effect was made in the Nationalization Act. In addition, the Board were empowered to disclaim contracts entered into shortly before nationalization if the contracts showed an 'unreasonable want of prudence'. The amending Act of 1949 empowered the Board to determine (on payment of compensation) a number of long-term contracts which hampered them in the efficient performance of their functions. The nature of the contracts affected by this provision is explained on page 131.

Various legal technicalities associated with the transfer of rights, duties and obligations, were solved by such devices as splitting contracts and adapting their terms to the new conditions, apportioning rents where properties were divided, and laying down rules as to the possession and right to production of deeds and documents. Most of the detailed provisions for these matters were outside the scope of an Act of Parliament and are to be found in the many regulations which have been made by the Minister of Fuel and Power under powers conferred by the Nationalization Act. Over forty sets of such regulations have been made and their total length is considerably greater than that of the principal Act.

One matter for which particular provision had to be made was liability for damage caused by subsidence resulting from mining operations prior to the date of nationalization. Subsidence of the surface may occur a considerable time after a seam of coal has been worked, and the normal legal rule is that a claim for damages does not accrue until subsidence takes place but is then enforceable against the person whose working caused the subsidence. In order to prevent colliery concerns from being indefinitely liable to subsidence claims it was provided, in effect, that the Board should be liable for damage occurring on or after January 1, 1947, although resulting from mining operations prior to that date. Where the damage occurred before January 1, 1947, but the resultant claim for damages had not been settled at that date, the Board were required to indemnify the colliery concern against expenses reasonably incurred in settling the claim. This provision as to transfer of indemnity was, of course, made subject to the Board having effective control over the conduct of litigation and the settlement of claims relating to subsidence damage.

*Compensation Provisions*

In providing for payment of compensation on nationalization a distinction is drawn between the value of assets for coal industry purposes and their value for subsidiary purposes, e.g. coke ovens, brickworks, or retail sales. Compensation for coal industry values will be paid out of a 'global' sum representing the over-all value of the coal industry. Compensation for subsidiary values will be paid separately on the basis of the market value of the 'compensation unit' comprising the assets concerned.

The global sum was fixed by a tribunal (known as the 'Greene Tribunal', after the chairman, Lord Greene) consisting of two judges of the Supreme Court and an accountant. The Government and the Mining Association had agreed that the value of the industry should be arrived at in this way, and the tribunal considered evidence submitted on behalf of the owners and of the Government. By adopting the principle of a global sum to represent the value of the coal industry as a single unit and as a going concern the Government avoided the inflated figure which would have resulted from adding together the values of the various items of property of which the industry was made up. The owners had the advantage, in the calculation of values, of the greater stability which the industry as a whole possessed as compared with any particular colliery. What the tribunal had to do was to ascertain the net annual revenue which the coal industry might reasonably be expected to earn in the future, and to fix the number of years' purchase which it was fair to apply in buying the industry. The value of the industry as determined by the tribunal was £164,660,000. This did not, of course, include the value of the minerals, for these had already been transferred to public ownership and £66,450,000 in compensation paid for them.

The task of ascertaining the extent to which assets are used for coal industry activities is facilitated by the fact that 'terms of settlement' (see page 18) between owners and mineworkers have, for many years, regulated the wages of mineworkers, and district practices have determined what assets of the coal industry are to be taken into account in ascertaining the wages to be paid; hence the expression 'T.S. assets'. T.S. assets which vest will be paid for out of the global sum; subsidiary assets will be paid for separately.

Compensation, separately ascertained, will also be paid for any increase of overhead expenses resulting from severance of industrial concerns. For example, if a colliery and a brickworks were in single ownership prior to nationalization, the overhead expenses of running the brickworks would almost certainly be increased if it was left with the owners when the colliery was nationalized. Provision was also made for the refund to colliery concerns of capital outlay incurred by them since August 1, 1945, with the approval of the Minister of Fuel and Power, in respect of any assets subsequently transferred to the Board. Without such a provision there would have been no inducement to owners to undertake capital development after the time when it was known that the industry would be nationalized and before transfer to public ownership took effect.

Twenty-one Valuation Districts have been established, corresponding in general with those for which district wages ascertainments were formerly made, and the global sum has been apportioned among these Districts by a Central Valuation Board. All transferred interests are in process of being made up into compensation units and allocated to the Valuation Districts. The compensation units will be valued by the appropriate District

Valuation Board, and disputed valuations will be reviewed by referees. The basis of valuation is a 'notional' sale of the compensation unit in the open market by a willing seller to a willing buyer. Compensation will be paid for the actual value so ascertained in respect of non-T.S. assets. The compensation for a unit consisting of T.S. assets will bear the same proportion to the District allocation of the global sum as the value of that unit bears to the total value of all the T.S. compensation units in that District. Thus, the total amount of the global sum will be exactly and fairly divided among all the owners of T.S. assets, though any one may receive more or less than the original valuation.

Compensation became due on the primary vesting date, but Parliament recognized that, because of the complexity of the task of valuation, several years would elapse before payment of compensation could be made. The Nationalization Act therefore provided for payment of interest, at rates to be fixed by the Treasury, on compensation from the primary vesting date until the date when compensation is satisfied. For the years 1947 and 1948 interest was fixed at 2½ per cent and 3 per cent respectively, but as the amount of compensation payable in respect of any particular colliery undertaking was not then known, revenue payments were made equal to one-half of the 'comparable ascertained revenue' of the undertaking during a typical year prior to nationalisation. A second amending Act of 1949 provided for the continuance of revenue payments for 1949 and subsequent years but at the rate of one-third of the 'comparable ascertained revenue' of the undertaking during the selected year. These revenue payments will be treated as advances on account of interest. Payment of compensation will be made in such a way as to safeguard the interests of mortgagees and others who may have interests in the property of a colliery concern.

Compensation will normally take the form of Government stock, but in some circumstances cash payments will be made, e.g. for colliery stock and stores. Interest on compensation and refunds of approved capital outlay will also be paid in cash. The Act of 1946 provided that compensation stock issued to companies was not to be sold without Treasury sanction, but this restriction was removed by the first amending Act of 1949.

The Nationalization Act contains many supplementary provisions, a number of which are examined in later chapters. It was necessary, for example, to provide fair treatment by way of compensation for people who had been employed in coal industry and certain other activities transferred to the Board and whose employment came to an end in consequence of nationalization, as well as to preserve pension and similar rights. The constitution of the Miners' Welfare Commission was amended so as to co-ordinate the activities of the Board and the Commission in matters relating to the health and welfare of Board employees, and various statutory selling schemes were wound up.

Chapter 3

# THE NEW ORGANIZATION

*Transition*

The Nationalization Act received the Royal Assent on July 12, 1946, and it remained for the Minister of Fuel and Power to fix the date on which the coal industry was to be transferred to public ownership. The decision was a difficult one. Government supporters looked for early fulfilment of one of the main features of the election programme, without fully appreciating the scale of the administrative problem which it involved. The mineworkers could not be expected, in the twilight of private control which they had long condemned, to give the new response which the industry so urgently needed. The colliery owners had little inducement to devote capital or energy to an enterprise which would no longer be profitable to them. These psychological factors had to be weighed against the administrative problem of bringing into being an organization capable of taking over the burden of the coal industry without collapsing under the strain. The Minister, after consulting the newly established Board, decided that it was better not to delay the change. Accordingly on November 18, 1946, he informed the House of Commons that he had decided that the primary vesting date should be six weeks later, on January 1, 1947.

In the meantime the Board had not been idle. Indeed, for some months prior to July prospective members of the Board had been meeting together as an Organizing Committee and making their plans. They knew that on nationalization taking effect the former companies would cease to operate the collieries and no organization except what they created would exist between the National Coal Board and the 1,400 mines up and down the country. There was no time to take the industry to pieces and refashion it according to some ideal plan, for the nation was in desperate need of coal and the Board would have to run the industry at the same time as they replanned it: at the same time, in fact, as they sorted out the vast miscellany of property transferred to them and made the acquaintance of their three-quarters of a million employees, who confidently looked forward to a new order. In the plans which they made the Board were assured of the good will of the unions representing the workers in the mining industry. Hardly less important in this time of transition was the fact that the colliery owners and the Mining Association co-operated fully in the transfer of the industry to public ownership.

The Board's administrative and technical staff were recruited from the coal industry, from officials of the mining unions and, to a limited extent, from outside the industry. The heavy demand for skilled personnel from within the industry created a difficulty, for a limited number of experienced men available were needed to build up the new organization at the same time as their services were essential to the running of the industry in the months prior to nationalization. This difficulty was overcome by working arrangements with the colliery companies and good will on the part of all concerned. Another problem which faced the Board, and more particularly the Divisional and Area Formations, in those early days was the provision of office accommodation. One of the solutions adopted was the purchase of large mansion houses, some of which were standing empty, and these purchases excited a certain amount of publicity and criticism. In fact these purchases were good bargains and the ability to house staffs in one building resulted in efficiency and economy in administration.

The continued functioning of the nationalized industry, notwithstanding its severance overnight from the boards of directors which had formerly managed its several parts, was ensured by the establishment of Coal Board Units under Controllers appointed by the Divisional Boards of the National Coal Board. These Units consisted of the staffs and general administrative services of the former colliery owners, but acting under the orders of the Controllers instead of the former owners or boards of directors. The Controllers were responsible to the Divisional Boards, or to the Board's Area General Managers where they had been appointed. Composite colliery and iron and steel works which had been controlled by a single company presented severance problems of particular difficulty, and in some cases the companies continued to manage the collieries on behalf of the Board. Arrangements had to be made, by means of temporary licences, to enable the owners of about 480 small mines (producing altogether nearly 2 million tons of coal a year) to continue to function pending a decision of the Board as to the best use which might be made of these mines.

A vast administrative problem was involved in preparing to assimilate the large number of colliery superannuation schemes and funds and in the establishment of the Board's own superannuation scheme. Nearly a thousand banking accounts had to be opened and put in funds in order that cash might be available for wages and other payments that had to be made in various parts of the country. Arrangements had to be made for splitting the accounts of many colliery concerns, for authorizing the making of payments and the issue and replenishment of stores, and for the necessary accounts to be kept for the recording of transactions. Some of the Board's financial officers were engaged in winding up the accounts of their former employers even as they established the new accounting systems of the Board. Decisions had to be made as to the vesting of assets subject to options, as to the carrying forward of cover for insurance risks, and as to the maintenance of statistical services. Negotiations on matters of high policy had to be carried on with the Government, with the Unions, and with commercial concerns and trading associations. Many problems of policy and of administration were still in course of solution, and many difficulties loomed ahead—some scarcely discernible—when the coal industry passed from private to public ownership on New Year's Day, 1947.

### *The New Order*

The new organization which the Board has established is based on the collieries. There are about 1,000 of these (excluding licensed mines), and prior to nationalization they were controlled by the boards of directors of more than 800 companies. Some grouping of collieries had been effected under private control but, as the Reid Report showed, the industry consisted of an excessive number of small units of insufficient capacity. It was imperative to reorganize the collieries in units of a size that would provide the maximum advantages of planned production.

The Board decided to group the collieries into 48 Areas under the control of Area General Managers, and to make the Areas the basic units for commercial management. The size of the Areas varies having regard to geological, geographical and other technical considerations; their number and distribution are not invariable, and some changes have already been made in the original lay-out. An Area may consist of from 6 to 66 collieries (the average number is 20) and have an annual output of coal varying from less than 1 million to more than 8 million tons.

The areas are grouped into nine Divisions based on the coalfields and follow, with some modifications, the regional system which the Minister of Fuel and Power established during the war. The Divisions are as follows:

| | |
|---|---|
| Scottish Division ... ... | Scottish coalfields |
| Durham Division ... ... | The Durham coalfield |
| Northern (Northumberland and Cumberland) Division | Coalfields of Northumberland and Cumberland |
| North Western Division ... | Lancashire and North Wales coalfields |
| North Eastern Division ... | Yorkshire coalfields |
| West Midlands Division ... | Black Country and Potteries coalfields |
| East Midlands Division ... | Nottingham, Derbyshire and Leicestershire coalfields |
| South Western Division ... | South Wales and coalfields of south-west England |
| South Eastern Division ... | Kent coalfield |

The South Eastern Division is small and constitutes in effect a single Area. It has no separate Area organization and lacks some of the features of the normal Divisional organization. A Divisional Board for each Division supervises and co-ordinates the work of the Areas in the Division, formulates Divisional policy, and is a means of decentralizing administrative control by the National Board.

The whole of the administrative organization between the collieries and the National Coal Board is the creation of the Board and not of Parliament. In law there is only one Board, and the

Nationalization Act contains no reference to Divisions, Areas or other subordinate formations. This is in marked contrast with the other nationalized industries. The Railway and other Executives associated with the British Transport Commission, and the Area Electricity Boards constituted under the British Electricity Authority, all exist as separate legal entities by virtue of statutory provisions creating them. The Area Gas Boards enjoy so great a measure of independence that the British Gas Council is little more than an advisory body. The National Coal Board, on the other hand, were left free to create such organization as they thought best fitted to serve the requirements of the industry. The Divisional Boards which they have created are not, for that reason, mere puppets moved by centrally operated controls; they have a virile life of their own and enjoy considerable autonomy. The National Board are, however, free to make radical changes in the organization if as the result of experience they consider such changes to be necessary. What they cannot do, under the existing law, is to divest themselves of the responsibility which Parliament has imposed upon them for the conduct of the industry.

The large scale of the Board's organization raises the difficult problem of how to distribute functions among those best qualified to handle them without at the same time dispersing responsibility for the decisions that have to be made. The system which the Board decided to adopt is that known as 'Line and Staff' and it is described as follows in the Annual Report of the Board for 1948:

> First there has to be a clear line of command from the National Coal Board to the coal face and of responsibility from the coalface to the National Coal Board.... The authorities in the 'Line' are the National Coal Board (collectively responsibile through the Minister of Fuel and Power and Parliament to the nation); the Divisional Boards (each collectively responsible to the National Coal Board; the Area General Managers (personally responsible to the Divisional Board); and the Colliery Managers (personally responsible to the Area General Managers).

An Area General Manager is assisted and advised by various Departments dealing with production, labour, finance, etc. These are his 'Staffs'. They may advise the Area General Manager (or the Colliery Managers) on technical matters, and he may delegate work to them, but he has undivided responsibility to the Divisional Board for what happens in his Area. Similarly at Headquarters and in the Divisions responsibility for action rests with the several Boards, and they are assisted by the specialist departments which constitute their 'Staffs'.

> All general instructions (the 1948 Report continues) pass between the authorities in the Line. For example, a decision by a Divisional Board that an Area should concentrate on preparing coal to meet the needs of a particular market would be communicated by the Board to the Area General Manager; guidance on the technique of coal preparation would be passed down the Staff channel from the Divisional Coal Preparation specialist to the Coal Preparation Engineer at the Area.

Authorities in the 'Line' are encouraged to exercise initiative, for their responsibility is clearly defined and the results of their actions may be readily judged. Their Staffs are intended to relieve them of the burden of a mass of technical and administrative detail.

*The National Headquarters*

The nature of the coal industry, and not merely the provisions of an Act of Parliament, make it necessary to have a central authority which can make policy decisions touching the industry as a whole. Someone must decide from a national standpoint the total amount of coal that is needed and how and where it is to be produced. This fact is temporarily obscured because shortage of coal makes it necessary for all the coalfields to produce as much coal as is economically possible; but sooner or later production will have to be allocated among the coalfields according to their ability to produce coal at reasonable cost and according to the needs of consumers at home and overseas. Coal must be marketed in accordance with national policy, for it would not be in the interests of the industry, nor in the long run of the country generally, for the coalfields to sell in uncontrolled competition with one another. The allocation of capital for development of the coalfields can best be made by a central authority which has regard to the needs and potentialities of all the coalfields and to general economic conditions. Similarly, broad principles of policy in regard to wages and conditions of service in the industry must be determined nationally.

For some years past the production and marketing of coal have been subject to regulation, and negotiations at a national level have taken place between representatives of Government, of colliery owners, and of mineworkers. The Coal Charges Account and earlier pooling systems, which struck a rough balance between profitable and losing collieries, may be compared with central control of the finances of the nationalized industry—necessary to ensure that it pays its way over an average of good and bad years. In fact the advantages of nationalizing the industry are realizable only through some form of central direction, and this direction is given by the National Coal Board.

The Board are responsible to the Minister of Fuel and Power, and through him to Parliament, for the efficient management and development of the coal industry and its associated activities. In turn the Board hold the Divisional Boards to account for the efficient discharge of the business assigned to them and to the administrative units subordinate to them. The Board formulate policy for the entire country in regard to production and development, labour matters, and coal sales. They ensure that as far as possible, training, research and welfare schemes match the needs of the industry as a whole. They provide technical assistance and advice, as well as common services appropriate to a national organization, for the benefit of the administrative and operational units which they have established. The Board also act as a central body representing the industry in negotiations with Government departments and trading and other associations. The Board's officers temporarily undertake on behalf of the Government certain functions of a regulatory character (e.g. the control of distribution of coal supplies) which depend for convenient or effective administration upon a central organization.

Originally the National Coal Board consisted of a chairman and deputy chairman, who were free of departmental duties, and seven 'functional' members who were heads of departments. Two members were at the head of the Production Department. The Marketing, Labour Relations, Manpower and Welfare, Finance, and Scientific Departments each had one Board member as head. Later, the Production and Marketing Departments ceased to be in the charge of members of the Board and the principle of part-time membership was adopted. In this way additional members of the Board became available to act, with the chairman and deputy chairman, as 'policy' rather than as 'functional' members.

At first the Board met twice every week, but later a General Purposes Committee with wide powers was appointed and this enabled the Board to hold less frequent meetings. The General Purposes Committee at first consisted of the full-time members of the Board who were not also heads of departments; other members attended only in an advisory capacity when matters concerning their departments were under consideration. Now all members of the Board are also members of the General Purposes Committee and take part in the decisions reached, but in practice the 'functional' members attend only for business which particularly concerns them. The Board have made extensive use of committees. Some of the more important, in addition to the General Purposes Committee, are concerned with Organization, Finance, Planning, Personnel, Wages Policy, Safety and Health, Education and Training, Output and Costs, Coal Preparation, and Coke. Representatives of the Divisions serve as members of many of these committees. The Board are also a constituent part of the Joint National Negotiating Committee, the National Consultative Council, and the National Miners' Welfare Joint Council (which are concerned with conciliation, consultation and welfare in the coal industry) and representatives of the Board serve on various other joint bodies which are representative of Government Departments, Trade Unions, and Trade and Technical or Scientific Associations. Close personal contact is maintained between Headquarters and the Divisions by holding regular conferences of chairmen, of administrators, and of technicians, and the benefit of Divisional knowledge and experience is also obtained by appointing men from the Divisions on the various committees and working parties which are set up.

A Director General, or official of equivalent rank, is appointed for each of the Headquarters departments, and the departments are divided into branches appropriate to the work undertaken. The Secretary to the Board, with the aid of a Principal Co-ordination Officer, co-ordinates the work of all the Departments, and is responsible for work not specifically assigned to other departments; in particular, establishments, administration of surface estates, statistical and economic intelligence, and public relations. The Legal Department, under the control of the Legal Adviser and Solicitor, advises all departments on matters of law, watches the Board's interests whenever legal considerations arise, and undertakes on behalf of the Board the work appropriate to a corporation solicitor.

*The Divisions*

For each of the nine Divisions a Divisional Board has been established very much on the lines of the National Board. The chairman and deputy chairman are free of departmental duties. Four other members (known as 'Directors') are responsible for Finance, Marketing, Production, and Labour respectively. (The two Headquarters departments of Labour Relations and Man-power and Welfare are fused into a single Department at Divisional level, and there is no member for scientific matters). Divisional Legal and Secretariat Departments are analogous to those at Headquarters. The original pattern of Divisional appointments was that the Directors were specialists in the particular Departments assigned to them; the Deputy Chairmen were men of wide experience in the coal industry; and the Chair-men came from outside the industry, being chosen for their sound judgment and wide administrative experience, particularly with a view to maintaining a balance of the various functions represented on the Divisional Boards. This pattern has altered somewhat in consequence of changes in the original personnel of the Boards.

The Divisional organization was created by the National Board in order to decentralize administration and take due account of the widely differing character of the coal industry in various parts of the country. The Board considered that if no adminis-trative unit existed between the 48 Areas and the National Headquarters the result would be either excessive centralization or diffusion of responsibility among a network of co-ordinating committees. The Divisional Boards have considerable freedom of action both in the application of national policy and in the formulation of their own local policies. They are allowed to incur capital expenditure up to £100,000 on any one colliery project without reference to Headquarters, and the 1948 Annual Report of the Board shows that capital expenditure amounting in that year to £27 million was authorized in the Divisions without reference to Headquarters.

The Divisional Boards furnish Headquarters with much of the information on which policy is based and they assist in the formation of national policy. They also formulate their own policies within the framework of national policy, co-ordinate the working and development of the Areas assigned to them, and provide advice and assistance to the Area officials in giving effect to those policies. The Divisional Boards provide various common services for the Areas, particularly in regard to marketing, financial, and legal matters. They represent the National Board in contacts with the regional formations of Government Departments and other national bodies, and negotiate matters affecting wages and conditions of service which are not of national implication. The Divisional Boards are also closely associated with the functioning of the consultative and welfare arrangements which have been established. The Divisional Boards have played a most active part in the formative period of the nationalized industry. With the development of Area organization and the steady reduction in the number of pits the function of the Divisional Boards to an increasing extent will be to advise and co-ordinate rather than to undertake executive management.

### *The Areas*

In each Area the Area General Manager acts as chief executive officer of the Board and is responsible for the management, reorganization and development of the collieries in his Area. Because of the imperative need for increased coal production and technical development, the Area General Manager was at first invariably a mining engineer and he was the chief production official in the Area. In that capacity he was responsible to the Production Director of the Divisional Board. In addition, so as to give coherence to the work of the Area, he acted as chairman of an Area Management Committee consisting, in addition to himself, of the Area officials responsible for labour, marketing, finance, science and administration. In this second capacity he was responsible to the Divisional Chairman for the efficient functioning of the Area as a whole.

This arrangement, imposing a dual responsibility upon the Area General Manager, was a provisional one and did not prove entirely satisfactory in practice. He was necessarily concerned with the technical details of production, and in addition he was deeply involved in administrative and co-ordinating functions associated with his office as chairman of the Area Management Committee. As the result of experience gained during the first year of nationalization the Board decided to appoint a Production Manager in each Area on the staff of the Area General Manager, and to make the Area General Manager fully responsible to the Divisional Board for the efficiency of the Area organization.

The Areas vary considerably in size and character, and as their organization is not required to conform to any uniform pattern local variations are found. Usually the Area General Manager is assisted by a Deputy, an Administrative Officer, and Area officials for production, labour, marketing, finance, scientific, and other matters. These Area officials have direct access to the Area General Manager and also to the Directors of the Divisional Board appropriate to their functions. Some of the larger Areas are split up and Sub-Area General Managers, Sub-Area Managers, or Agents are responsible to the Area General Manager for the operation of groups of collieries within the Area.

An Area is a commercial undertaking of considerable size and importance—the average output being 4 million tons of coal a year with an annual turnover of about £10 million—and its successful operation depends upon the ability and personality of the Area General Manager. The appointment of Area General

Manager is one of the most important in the Board's service, calling for exceptionally high standards of administrative ability, sound business sense, and understanding of the special difficulties of the mining industry.

*The Collieries*

The Colliery Manager occupies, in relation to a colliery, a position somewhat similar to that of the captain of a ship. He is responsible for the effective operation of the mine, and the safety of hundreds, perhaps thousands, of men depends upon his skill and judgment. His personality and ability to handle men can make all the difference between an efficient, contented pit and one in which disputes and grievances are constantly recurring.

The Coal Mines Act 1911 requires that every mine (except a few small ones) shall be under one Manager who shall be responsible for its control, management and direction, and daily supervision. No one who is not qualified to be a Manager may take part in the technical management of a mine. The number of mines which may be under the control of a single manager is strictly limited.

The Manager acts in accordance with the instructions and advice of the Area General Manager and his Agent or Sub-Manager, but he is personally responsible for strict compliance with the provisions of statutes and regulations relating to safety and health in mines. His duties are too numerous to be detailed here. In general they comprise over-all responsibility for coal face operations, support, ventilation and lighting, haulage and signalling systems, shafts and winding, training and working arrangements generally above and below ground, and observance of proper precautions in the use of machinery, electricity and explosives. It is the duty of the Colliery Manager to withdraw the men from a mine or part of a mine which is found to be dangerous. The Manager gives effect to Board policy in the operation of the mine and he has important duties in connection with its planning and development. He has to make various reports and returns, and to assimilate and apply the instructions and guidance which he receives from the authorities above him. He is closely associated with conciliation, consultation and welfare activities associated with the colliery for which he is responsible.

The Manager may be assisted by one or more Under Managers, and they must hold at least a second-class certificate of competency. In addition to the general assistance which they give to the Manager, Under Managers' duties are to enforce the statutory provisions as to safety and health, to investigate complaints, to confer daily with the other underground officials as to the working of the mine, and to see that all materials and applicances are available as required. Often it is the Under Managers who have direct contact with the men and who engage or dismiss them. Still in the 'line' of command are the Overmen—underground officials who assist the Under Managers by performing duties somewhat comparable with those of foremen in surface jobs. Firemen, Examiners and Deputies (the terms are synonymous) with approved qualifications are appointed, in compliance with the 1911 Act, to make inspections and carry out such other duties as are necessary to detect the presence of gas and supervise the ventilation, state of roof and sides, and general safety of the part of the mine to which they are assigned.

Mine Managers are assisted by technicians, including engineers, surveyors, training officers and safety officers. The general working of a colliery calls for the services, in addition to the miners, of a great variety of skilled and unskilled labour, including electricians, fitters, lamproom men, horsekeepers, weighmen, coal preparation and pitheap operatives, storemen, cashiers and clerks.

*The Burrows Committee*

In May 1948 the Board appointed a Committee consisting of Sir Robert Burrows (recently appointed a member of the Board) as chairman, and two industrialists, from outside the coal industry, and requested them to make a thorough review of the Board's organization. The Board published the main recommendations of the Committee together with their own observations on them.

The Committee and the Board were in agreement as to the principles which should govern the establishment and functioning of a large organization, viz: 'as much devolution of responsibility as possible; the creation of a clear channel of command; and the maintenance of a clear distinction between policy-making and execution, but not so as to prevent the association with policy-making of those responsible for execution'. They were principles to which the Board were giving effect in their organization. The Committee recognized that the problem facing the Board was more difficult than that which faced even the largest-scale undertakings under private enterprise, since those undertakings grew gradually whereas the Board's organization had to be created suddenly. A sound scheme of organization, the Committee remarked, did not in itself guarantee efficiency; that would only come gradually as those engaged in management acquired the habit of working together, and as suitable people were found or trained for managerial appointments.

Provision has now been made by Act of Parliament (see page 38) for giving effect to the recommendation that the Board should be enlarged by the appointment of a second deputy chairman and part-time members. The Board agreed with the Burrows Committee that members who are in charge of functional Departments should leave executive action to their chief officials although retaining responsibility for their Departments. They did not accept a recommendation that Divisional Chairmen should serve as members of the National Board, for they considered that the objections to such an arrangement outweighed the advantages. They thought that it would confuse the clear channel of command between Headquarters and Divisions and might result in conflicting loyalties where National and Divisional interests diverged. They also thought the arrangement would distract Divisional Chairmen in the carrying out of their very considerable duties in the Divisions. The Board have already given effect to recommendations of the Committee for the increased use of committees so as to lighten the load on the National Board and for associating Divisions with policy formation and national negotiations on wages and salaries.

The interlocking structure which the Burrows Committee had in mind is reflected in their suggestion that Area General Managers should be members of Divisional Boards. The Board were not prepared to adopt this suggestion, again because it conflicted with the fundamental principle of maintaining a clear channel of command between Headquarters, Divisions, Areas and Collieries. They also considered that the arrangement would make the Divisional Boards too large to be efficient. They agreed as to the value of part-time membership of such Boards. A recommendation that the position of Colliery Managers should be strengthened reflected the feeling that Managers did not enjoy the same authority in the service of the Board as they had done prior to nationalization. The Board declared their intention to uphold the prestige and authority of the Managers, but emphasized the necessity, in a nationalized industry, for such matters as the fixing and variation of price lists and the settlement of disputes to be co-ordinated at Area or Divisional levels.

Chapter 4

# HOW COAL IS MINED

*Coal Formation and Resources*

The coal in Great Britain was formed many millions of years ago by the action of heat and pressure on decayed vegetation. Giant fern-like trees grew in and near swampy lakes and, as they decayed, formed deposits of vegetable matter. Changes in the surface of the earth resulted in these deposits being covered by layers of sand and mud and succeeded by further layers of vegetation and sediment. Later these layers became buried under hundreds or even thousands of feet of rock, and the heat and enormous pressure of the rock changed the vegetable matter into coal. The muddy sediments became the shale and other dirt now found with the coal. Earth movements resulted in the layers, or strata, of rock and coal being bent and folded, so that sometimes the coal outcrops to the surface, and at other times lies thousands of feet below. The strata may have been fractured by these movements, resulting in faults as shown in Fig. 1. The coal occurs in bands from a few inches to several feet in thickness and varies considerably in chemical composition and quality. The distribution of the coalfields is shown on the frontispiece map.

The presence of coal may be apparent from the fact that the seam is exposed at the surface ('outcropping') or may be ascertained by boring. In those parts of Great Britain, particularly the north and west, where the surface rock is older than the carboniferous period when the coal measures in this country were formed, there is no possibility of coal underlying the rock. In the South Midlands and Southern England, where the surface rock was formed subsequent to the carboniferous period, there is every possibility of new coalfields being discovered. An example of comparatively recent discovery is the Kent coalfield, which was not proved until 1890 and now yields over a million tons of good coal a year. It is estimated that the British coalfields contain reserves of coal capable of affording supplies for about two centuries at the present rate of production. Seams of coal less than 18 inches thick are in the main not a workable proposition, and coal lying more than about 4,000 feet below the surface is not usually worked because of the high temperature of the strata (about 110° F.) at that depth.

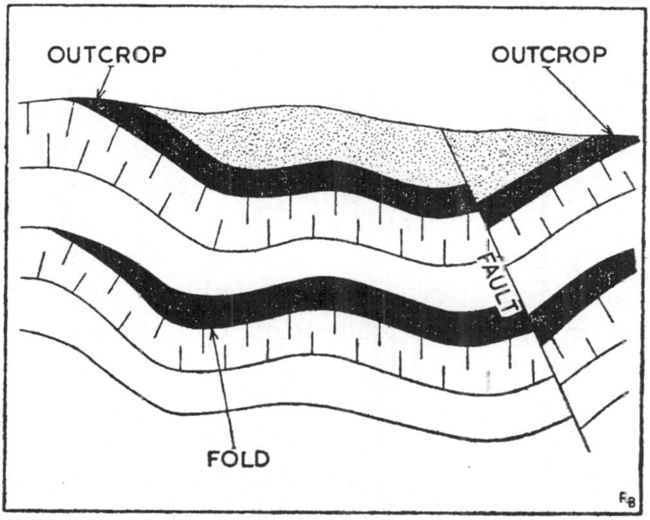


Fig. 1. Strata with seams of coal

Apart from opencast working and drift mining, which are referred to later, coal can only be worked by sinking shafts from the surface and driving headings into the coal. When the geologist has located coal the mining engineer is faced with a number of problems, the principal of which are: reaching the coal with workings; extracting the coal; lighting, ventilating and draining the workings; controlling subsidence; transporting men and materials; and disposing of the 'spoil', or waste material, resulting from mining operations. The cost of opening a new mine may amount to several million pounds, and before embarking on such a venture it is essential to have some idea of what it will cost to establish and operate the mine, and what the yield of coal is likely to be. Material extracted in the process of boring will enable the geologist to tell the depth and thickness of the coal seams, how many seams there are, whether they are horizontal or inclined, what quality of coal they contain, and the nature of the strata through which shafts will have to be sunk. With the help of this information the mining engineer is able to plan the mine which will be necessary to work the coal economically and efficiently. Financial and marketing experts will calculate whether the yield from the mine will make an adequate return on the capital which it will be necessary to invest in it.

*Shafts and Roadways*

For every separate mine two shafts are necessary for ventilation. One is the 'downcast' shaft, and through it fresh air is drawn for the purpose of ventilating the underground workings. The air returns to the surface by way of the 'upcast' shaft. The shafts are normally circular and may extend to a depth of 4,000 feet. Sinking them is a major engineering task that may take several years and present considerable practical difficulties, particularly where the shafts have to be driven through sand or water-bearing strata. Sometimes the intrusion of water is prevented by freezing the ground through which the shaft is being sunk and maintaining a collar of ice around the shaft workings until a waterproof lining has been completed. All labour and materials as well as drainage pipes and power lines, passing into and out of the mine must do so by way of the shafts, and the maximum possible output of a mine is largely determined by the capacity of the shafts. The shafts may easily become bottle-necks, and once they are sunk and the mine is in operation it is practically impossible, without closing the mine down, to increase the diameter of the shafts in order to cope with a greater output.

In each shaft there are two cages with separate ropes attached to them but connected to a common winding drum. Powerful winding engines move the cages at great speed in opposite directions, so that one cage is ascending as the other descends. At the beginning and end of a shift men are taken to and from the surface; at intermediate times the cages are used for raising coal or waste materials, or taking plant and materials into the pit. Skip-winding is a means of improving shaft capacity in raising coal; the skips are large hoppers which are automatically filled with coal at the bottom of the shaft and automatically discharged at the surface. The system is very economical in the use of labour and enables maximum use to be made of shaft capacity.

The shafts must rest upon a firm foundation and so it is not possible to mine the coal in the immediate vicinity of the shaft bottom—otherwise subsidence would result in serious damage to the shafts. The shaft bottom is like a busy railway junction, with trucks laden with coal or debris constantly arriving and being sent to the surface, whilst empty trucks and materials are being sent into the mine. A well-planned shaft bottom will make an enormous difference to the efficiency and economy of labour in maintaining the movement of full and empty trucks.

From the shaft bottom roads are driven 'inbye'—that is, in the direction of the coal workings—and in the course of time these roads may extend several miles underground. In a modern lay-out the roads are usually constructed in sets of three, one being a travelling road for the miners, one a haulage road for coal, and the other a road by which foul air is extracted from the mine.

If the underground roadways could be driven straight and level through coal, conditions would be ideal, but this is seldom possible, because of the nature of the strata, and usually in this country the roadways follow the rise or dip of the seam. Heavy gradients resulting from following an inclined seam make it impossible to take advantage of locomotive haulage underground. The Continental practice is to drive level roads through the strata; this is more expensive initially than driving through coal but may pay in the long run because of the improved haulage arrangements that are possible. A saving is also effected in 'ripping' and the repair of roadways, which is usually very costly when the roadways are driven in the seam. 'Ripping' consists of cutting away the rock above the seam so as to increase the size of the roadway. Before existing British mines can be equipped with locomotive haulage it will be necessary for their lay-out to be fundamentally changed and a great deal of money spent on driving horizontal roads in stone.

### *Extracting the Coal*

The oldest system of mining coal, and one used extensively in the United States is known as 'room (or bord) and pillar'. In this system narrow headings are driven into the coal seam in such a way as to divide it up like a chess-board, as shown in Fig. 2.

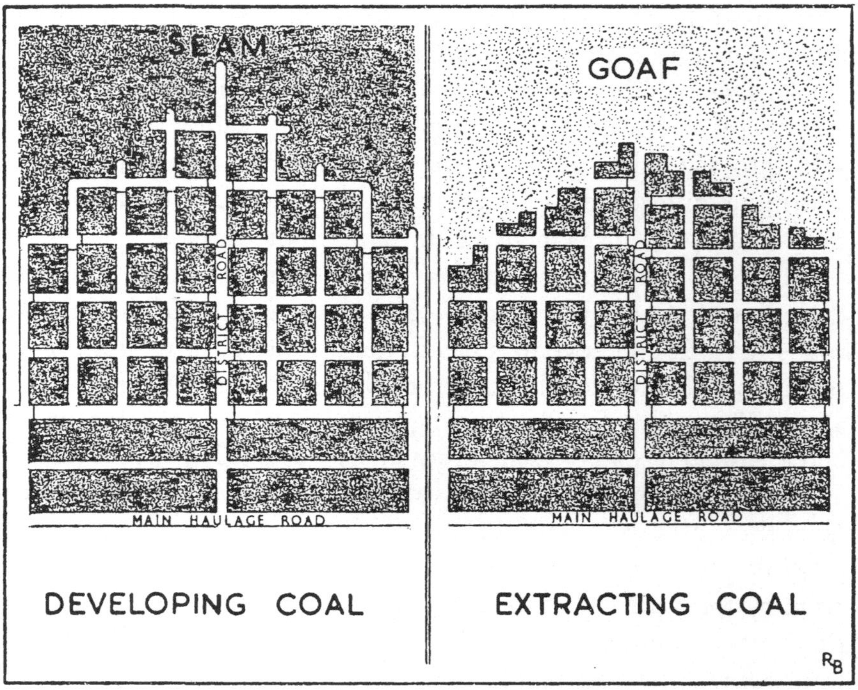


Fig. 2. Room and Pillar Working

The pillars are the squares of the chess-board, and may have an area of from 400 to 4,000 square yards; they are not necessarily square. When the working area has been cut into pillars the pillars may in turn be removed and the roof allowed (under control) to break down into the 'goaf' or space from which the coal has been removed. Geological conditions including depth of seam normally determine whether the room and pillar system is the best to use in any particular mine. It is economical in labour, utilizes the services of men working in small skilled groups, and is free of the rigid cycle of operations which other systems impose. The system involves greater technical difficulty than 'longwall' mining and therefore requires more skilled planning and technical staff;

capital expenditure is relatively higher and the percentage of extraction of coal is somewhat less. Control of subsidence is also more difficult with the room and pillar system.

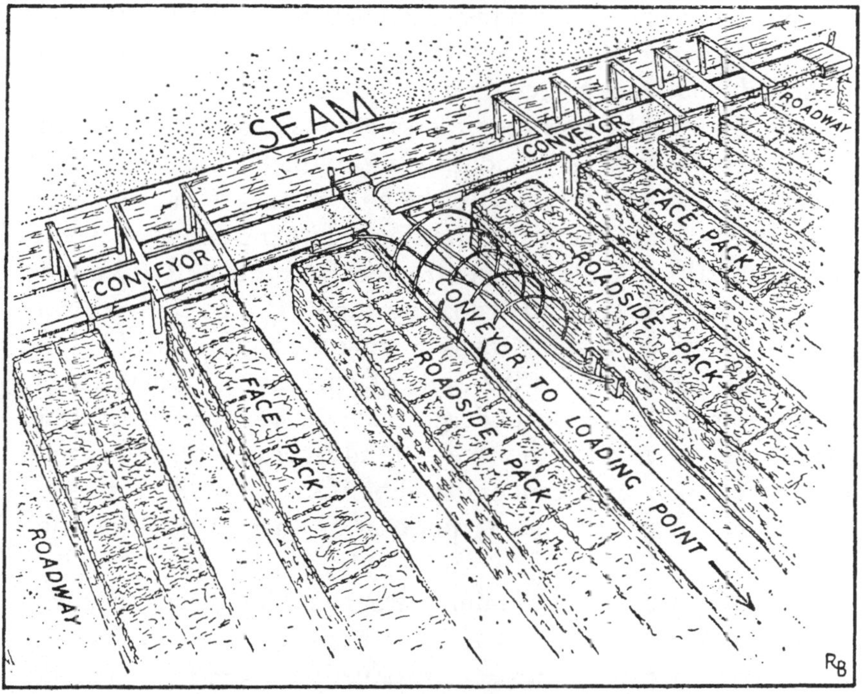


Fig. 3. The Longwall System

The longwall system is used in most British pits. Instead of cutting the coal up into a number of relatively small pillars the coal is worked along a face that may be several hundred yards in length. As the face advances the space from which the coal has been removed is packed with stone so as to control subsidence, and roadways are extended and supported in order to serve the working face. Conveyor belts in one or more sections run parallel with the face, and by a series of 'T' junctions convey the coal to the loading point, as shown in Fig. 3.

Efficient operation of the longwall system necessitates strict observance of a 24- or 48-hour cycle of operations. Local practice varies as to the shifts on which particular operations are performed, and some operations may be performed on more than one shift. The following is a typical cycle of operations. On the night shift a section of coal the length of the working face is undercut by machine to a depth of 4 or 5 feet into the seam. The morning shift breaks down the coal which has been undercut, sets props to prevent the roof from collapsing where the coal has been removed, and loads the coal on to the conveyor. The weight of the undercut coal assists the process of breaking up the coal, but explosives (shot-firing) and pneumatic picks are used when necessary in order to break up the coal so that it can be loaded on to the conveyor. On the afternoon shift the conveyor is dismantled and re-erected nearer to the face. Packs are built to control the collapse of the roof and supports are removed from the line of earlier working thus allowing the roof to subside. The night shift recommences the cycle of operations. On the afternoon and night shifts the underground roadways are advanced in order to keep up with the face as it advances. With the system described about 50 per cent of the labour force would be employed on the morning shift and 25 per cent on each of the other two shifts.

Advantages of longwall mining are the high percentage of coal extraction which it makes possible, the small amount of development work necessary prior to productive operations, and the control of subsidence which is possible. Disadvantages are the amount of labour and material required for support of roadways and at the working face, and the fact that, because factory methods of supervision cannot be applied to mining, lower standards of work and discipline are sometimes associated with the larger concentrations of men that the system requires. The rigid cycle of operations necessary for longwall working is easily disorganized by absenteeism and slow working.

The longwall system may be either 'advancing', i.e. working away from the shaft, or 'retreating', i.e. driving headings into the coal and then working back towards the shaft on a longwall face. An advantage of longwall retreating is that in driving the headings a good deal is learnt about the nature of the seam to be worked, and so the workings can be more accurately planned.

Moreover, expenditure of labour and materials on road maintenance is sometimes reduced because the roadways have support from the coal. The principal disadvantage of longwall

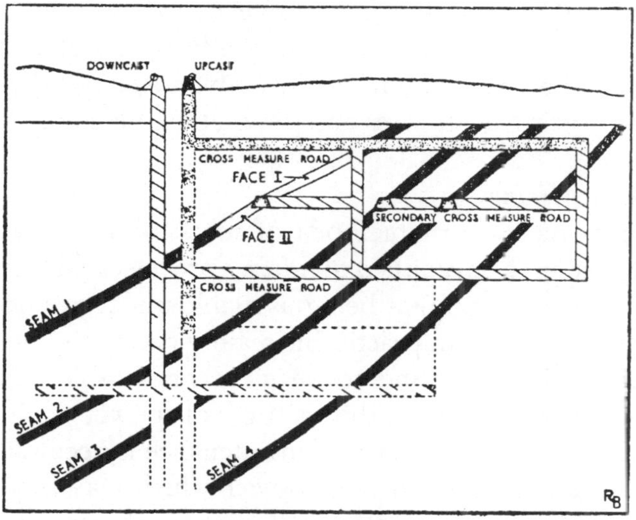


Fig. 4. Horizon mining

retreating is high cost and low output during the slow process of driving the initial headings, but this should not be serious now that sufficient capital and suitable machinery are available. The 'Reid' Report recommended that for the attainment of highest productivity the intensive room and pillar system should receive first choice. The longwall retreating system was recommended as second choice as it combined the advantages of the other two systems while avoiding some of their disadvantages.

In planning new mines the Continental practice of 'horizon' mining will be followed wherever suitable. From the shaft bottom the main intake road is driven level through the strata so as to intersect the inclined seam at a predetermined point (the 'horizon'); the road may in fact have a slight gradient in favour of the load. At a higher level the main return road is driven from the upcast shaft also to intersect the seam, as shown in Fig 4.

The section of seam so interesected is then worked, usually by longwall advancing, in suitable units. Working roads driven in the seam lead up to the main return road and are connected by short vertical shafts to the main intake road lying below. As the main haulage roads are level they are suitable for locomotive working; at the same time the difference in levels between the intake and return airways gives the advantage of ascensional ventilation (i.e. ventilation taking advantage of the normal tendency of warm air to rise) and gravitational loading.

*Mechanized Mining*

In recent years modern methods of mechanization have to an increasing extent been applied to mining processes. Mechanization necessitates power supplies being available underground and the use of steam power is impracticable mainly because of heat losses. Compressed air was found to be more practicable and is now used extensively, though this source of power is inherently inefficient and a high standard of maintenance is necessary. Compressed air is the only source of power which may be used near the coal face in gassy pits; electricity may not be used because of the danger of explosions.

Electricity is by far the most efficient form of power supply when it can safely be used underground. Transmission losses are relatively small, pressure of supply can easily be maintained under working loads, and the supply lines are more manageable than the rigid pipe systems necessary with compressed air. The disadvantages of using electricity are the danger of fire and of gas in the mine being ignited by a spark from defective electrical equipment or from a fractured cable. Much has been done in the way of designing flameproof electrical apparatus capable of being used with safety underground, but the effective answer to the problem is improved underground ventilation which will make the pit safe for the apparatus, rather than to try and make the apparatus safe for the pit. There is, however, room for considerable improvement and standardization in the design of electrical apparatus for underground use.

Almost 20 per cent of the output of coal in this country is still obtained by hand-wielded picks. When the coal is soft improved output may result from the use of pneumatic picks operated by compressed air—rather on the lines of pneumatic road drills. Shot-firing for the purpose of breaking down the coal is then practically eliminated and getting the coal is still sufficiently an individual task for dirt bands to be avoided. The use of mechanical coal-cutters is steadily increasing and these powerful machines cut or shear the coal from the face, the coal then being blasted or broken down and loaded by hand. Complete mechanization of face working is reached with machines which both cut and load the coal; mechanical loaders are also being developed which load the coal after it has been hewn by some other method. Under suitable conditions mechanized working at the coal face will result in considerably increased OMS (output per man shift) but a high degree of mechanization is not free from disadvantages. Machines are less selective than individual labour, and in consequence a good deal of dirt is produced with the coal. They also create much dust and noise in operation and the speed and power with which they work increase the difficulty of roof control.

In some pits the tubs of hewn coal are still hauled from the working face by human or pony labour, but conveyor-belt systems are now used to an increasing extent both at the face and in the subsidiary haulage roads. In the main haulage roads the tubs of coal are hauled to the shaft bottom by endless rope haulage systems driven by electricity or compressed-air motors. The most efficient method of main-road haulage both for men and materials is by diesel or electric locomotive, but few pits in this country are designed for such working. Complete overhaul and re-development of mechanized underground transport offers one of the best means of improving productive standards in the nation's pits. There is considerable need in particular for man-riding facilities underground. In some pits the distances which men have to walk from the shaft bottom are so great that the actual working time at the coal face may be not more than 5 hours in a shift of 8 hours.

### *Ventilation, Lighting and Drainage*

An efficient system of ventilation in a coal-mine is essential for safety and efficiency. Methane gas, known as 'firedamp', is given off by the coal seams and a concentration of 5 to 15 per cent in the air forms a highly explosive mixture. Pits and individual seams vary in the amount of firedamp given off, but in a 'gassy' pit very stringent precautions are taken to prevent the introduction, accidental or otherwise, of naked lights. Smoking is prohibited and pipes, cigarettes, matches and lighters are 'contraband' which may not be taken into the pit. Electricity for power or general use is prohibited or subject to severe restrictions, and handlamps or caplamps, whether oil or electric, must be of approved pattern and properly maintained. Some pits, however, are gas free and known as 'naked light' pits.

Coal dust suspended in the air forms an explosive mixture and frequently roadways have to be dusted with a particular type of stone dust in order to reduce coal dust concentrations. Both coal dust and certain types of stone dust in the air are known to be capable of causing serious lung diseases, such as pneumoconiosis, by the dust particles being breathed into the lungs, and by spraying water and other wetting agents the amount of dust in the air is reduced. Quite apart from safety requirements, the health and comfort of the miners and their efficient working necessitates a good standard of ventilation, particularly in the deep and hot mines.

Originally mines were ventilated by a draught created by a fire burning at the bottom of the upcast shaft. This method is no longer in general use; instead, large fans draw the stale air out of the mine, or sometimes force the air through it, often at a rate approaching half-a-million cubic feet a minute. Underground it is essential to ensure proper circulation of the fresh air throughout the workings and to avoid short circuits which will result in particular sections of the mine not being ventilated. This is achieved by a carefully designed system of airways, with air-lock doors and curtains in order to prevent leakage from the intake to the return airways. With ventilation, as with so many other features of coal-mining, radical improvement is often only possible by the complete re-development of the mine, but improvements are possible by careful surveys and overhaul of existing ventilation systems.

Good lighting in mines is difficult to achieve, to some extent because of restrictions on the use of electricity, but chiefly because the nature of coal face operations involves considerable risk of damage to electrical equipment and necessitates the continual dismantling and re-erection of apparatus. As a mine may have 50 miles or more of roadway in use universal mains lighting of roadways is impracticable. In consequence lighting is supplied by electric or oil lamps which the miner wears on his helmet or carries in his hand (in some mines candles are used), and this lighting is not really adequate for the performance of dangerous and difficult tasks. Low standards of lighting increase the risk of accident, reduce working efficiency, and are considered to be a contributory factor in causing a disease of the eyes known as 'miner's nystagmus'. Work is being undertaken in the improvement of lighting underground, particularly in the use of fluorescent lighting at the coal face.

In the nature of things a mine is likely to accumulate a good deal of water and some pits produce, ton for ton, much more water than coal. This water has to be pumped from the mine by way of the shafts, and even though it may be disposed of at the rate of a thousand gallons or more a minute, working conditions in a wet mine can be extremely unpleasant.

*Subsidence*

Underground workings, whether in stone or in the coal seam, are subject to very considerable pressure, varying according to the depth of the seam but amounting in deep mines to as much as 4,000 lbs. to the square inch. This pressure is not simply a dead weight from above but is exerted in all directions, so that in an excavation there is a tendency for the roof and sides to collapse and for the floor to buckle. As coal is extracted, wood or metal pit props are set in order to support the roof at the working face, and the systematic placing and removal of props forming the face supports is essential for good roof control and for preventing undue disturbance of the strata. As the working face advances the roof may be allowed to cave in over the 'goaf' or it may be supported by packs of stone, built like walls at right angles to the face.

Alternatively the goaf may be packed solid with waste material. Hydraulic or pneumatic methods result in good solid packs which minimize subsidence, improve ventilation (by preventing air leakages) and reduce the chance of goaf fires. These methods are, however, expensive to operate, especially where sufficient suitable waste material does not become available in the process of mining.

Fractures and movements in the strata cause a squeezing in of the roadways, and much labour has to be employed in ripping and repairing the roof as well as the sides and floor of roadways. Altogether there are about 15,000 miles of underground roadway much of which requires continuous support. This is provided by means of timbering and steel arches, and by the building of roadside packs which deflect the main pressure away from the road itself.

The subsidence which takes place in the mine eventually affects the overlying surface land, with consequences that may be serious. Unless preventive action is taken, subsidence at the surface amounts to about one-half of the thickness of coal removed, and will extend to a distance beyond the vertical line above the working face equal to about a quarter of the depth of the seam being worked. In the area 'let down', drainage systems of agricultural land may be destroyed, highways and public service pipes fractured, and surface buildings suffer anything from a few cracks to complete destruction; it does not, however, necessarily follow that extensive damage will be caused. Land not immediately above the underground workings may be affected by what is known as 'draw'; that is, the tendency for strata to be drawn in towards the point of subsidence. Damage resulting from subsidence can be considerably reduced by adopting suitable mining methods. The longwall system of mining which advances fairly rapidly on a broad front causes less surface damage than is the case with the room and pillar system of working. Where solid stowing is economically possible the extent of subsidence damage is further reduced. In addition, structural precautions can be taken in surface building in a mining area so as to prevent or minimize damage from the workings. Pillars of coal may be left for the support of buildings of historic, aesthetic or industrial importance, or which provide some essential service, but these pillars have the disadvantage of causing distortion at the junction of the supported and unsupported surface land.

The law as to support from minerals underlying the surface is extremely complicated. In the past land-owners used to grant someone the right to work coal but retained the surface land themselves, or they worked the coal themselves and sold the surface land. In consequence the working of the coal became subject to rights and obligations varying from complete freedom from liability for surface damage, to full liability for the consequences of such damage. In addition, railway companies and authorities providing public services such as water, drainage and lighting, have had special and often inconsistent statutory codes applied to them. Problems associated with mining subsidence were considered by a Committee appointed by the Minister of Fuel and Power, which reported in 1949. The Committee recommended that in future a right of support for surface land should be acquired or retained only when the national interest so demanded, and that compensation should be provided for all surface interests which suffered damage from subsidence caused by working coal. Certain key points, such as power stations, and reservoirs, as well as buildings of historic or aesthetic interest would, however, be entitled to a statutory right of support. Special provision would be available for railways, canals, highways, and land requiring special drainage provision. The Committee also recommended a simplified procedure for dealing with claims to compensation for subsidence damage. As the recommendations would substantially increase the National Coal Board's liability for subsidence damage it was proposed that the Treasury should make an annual grant to the Board in respect of the additional cost. The Committee emphasized the fact that town-planning schemes should minimize the amount of damage resulting from mining subsidence by co-ordinating underground and surface development.

*Coal Preparation*

The coal which is brought to the surface usually requires some form of preparation before it is fit to market. In particular, the dirt or stone which becomes mixed with the coal in the process of mining should be removed, and the coal graded into suitable sizes. In recent years there has been an increase in the amount of dirt in coal reaching the consumer and there are a number of reasons which account for this. One is that the thicker and cleaner seams have been very largely worked out in many of the existing mines, and in consequence there has been a slow decline in quality over a period of years as it has become necessary to work seams which contain dirt bands. Another reason is that in the immediate post-war years there has been an imperative demand for coal, and in the attempt to meet the demand quality has had to be sacrificed to some extent in favour of quantity. This factor is closely linked with the development of mechanized production. The machine does not discriminate in the same way that is possible when coal is handgotten, and in consequence additional dirt is brought to the surface with the coal. The remedy lies in providing additional coal preparation plant. Up to 1939 the installation of such plant did not keep pace with mechanized production, and ceased almost completely during and just after the war. The amount of dirty coal distributed is also affected by the fact that 5 per cent of the total output of coal is from opencast workings. This coal tends to be dirtier than deep-mined coal. It is not practicable to provide preparation plant at the short-lived opencast sites, and it is an expensive process to convey opencast coal to a colliery washery for treatment.

To some extent public concern over dirty coal is related to price. During the war flat-rate increases in the price of coal resulted in low quality coal costing proportionately much more than the better quality coal. People are less concerned if they get a low quality coal at a low price, and the revised price structure which the Board intend to introduce (see page 139) should, in the long run, remove this sense of grievance among consumers.

The simplest form of cleaning coal is by hand picking and this process is carried out as the coal moves along travelling belts. Mechanical cleaning consists of 'washing' the coal, and takes advantage of the fact that the specific gravity of coal is less than that of the dirt mixed with it. Coal can be 'floated off' in a liquid of appropriate specific gravity, leaving the dirt to sink as a residue. Dirt embedded in the coal is less easily dealt with, but efficient coal washing machinery will separate out 'middlings'. Middlings usually consist of pieces of coal with shale adhering to them. If they are crushed the pure coal is released and can be recovered by re-washing. For some purposes it is disadvantageous to have wet coal, and dry cleaning is then resorted to instead of 'wet' washing. The finest sizes of coal cannot be cleaned by a 'sink or float' process depending upon difference in specific gravity, and instead 'froth flotation' is used. This process depends upon the fact that oil has a greater 'affinity' for coal than for dirt. Coal and dirt are immersed in water to which a small quantity of oil has been added and the oil adheres to the coal in preference to the dirt; bubbles generated by a froth flotation plant then attach themselves to the coal particles causing them to float, but the dirt particles, not being so supported, sink, and so the coal is separated from the dirt.

Although coal cleaning is an expensive process it is normally well worth undertaking because of the waste involved in transporting and distributing dirt, which is of no value in itself and may result in considerable output loss when fed as fuel into boilers, and may also create problems of ash disposal. In addition, coals used for the production of blast-furnace and foundry coke must be low in ash content in order to minimize loss of iron in the furnace slag; coal for gas-works must be clean in order to yield good, clean coke.

At the pithead, coal is sorted for size by passing it over a series of screens. These screens have apertures varying in size between one screen and another. They are mechanically shaken so that the coal is jerked along and falls through the apertures of the screens into compartments appropriate to the various sizes of coal. In recent years an attempt has been made to standardize the sizing of coal into the grades shown in the table on page 128, but it is not possible within a short time to replace the many thousands of screens in use. Some tolerance is necessary in the standard sizes, as marketing considerations enter into the proportions in which the various sizes are produced; these proportions are capable of variation by altering the upper and lower limit of the standard sizes.

A particular problem at the pithead is the disposal of the waste material necessarily brought to the surface in the course of mining operations, and resulting in the unsightly pit heaps which disfigure the countryside. It seems appropriate that this waste material should be returned to the mine, but the cost of doing so is usually prohibitive, and it would not be feasible to divert sufficient labour to the work. Town-planning control should in future ensure that pit heaps are sited where they are least objectionable, and the encouragement of bushes and other natural growth on old pit heaps will make them less unsightly. Some pit heaps are liable to spontaneous combustion, and one tip has been burning for several decades, all attempts to extinguish it having failed. Efficient coal preparation will reduce the amount of combustible material in pit heaps and so reduce the chance of such fires occurring.

*Coal Products*

The most important and best known products of coal are coke and gas. They are produced by a process known as 'carbonization', in which coal is heated in retorts and ovens until gases are given off, leaving coke as a residue. High temperature carbonization, at a temperature of 1,000° C. or more, is used more particularly for the cokes required for metallurgical purposes, but also for coke for domestic and other uses. Low temperature carbonization, at about 650° C., is used for the production of domestic coke for use in open fires. It is most important for the efficient operation of carbonization plant to use suitable and uniform qualities and grades of coal, and the scientific grading of coal referred to on page 100 is of the greatest assistance to the gas and coking industry. The National Coal Board took over a large number of coking plants and produce more than two-fifths of the national output of hard coke—the oven coke used primarily for blast furnaces in the iron and steel industry.

The gas which is given off in the process of carbonization is treated in order to extract tar liquor, sulphate of ammonia, benzole, and numerous derivatives, generally known as coal by-products. After this treatment the gas becomes the familiar 'town' gas used for domestic and other purposes. About 24 million tons of coal are used every year in gasworks for the production of town gas; the National Coal Board produce about 8 per cent of the nation's supply of such gas.

Experiments have been made here and in other countries with a process known as the underground gasification of coal. This involves setting fire to the coal in the seam underground. The gas produced by the consequential total gasification is led to the surface, purified, and consumed in gas engines which drive electricity generators. In this way easily distributable power is produced without the necessity of hewing coal, bringing it to the surface, and cleaning and distributing it. Much practical experience in controlling fires in the seam will be necessary before underground gasification passes beyond its present experimental stage.

The primary by-products of coal are tar, benzole and sulphate of ammonia, the latter being a valuable fertilizer. The Board produce a considerable part of the national output of these commodities, as well as such derivatives as pitch, naphthalene, and a wide range of raw materials used in chemistry and in the paint and dye and plastics industries. Analogous to coal by-products are briquettes and manufactured fuels which make use of coal that might otherwise be wasted. Briquettes are made by mixing fine coal with pitch and pressing the mixture into moulds. A popular smokeless fuel known as 'Phurnacite' is made by 'briquetting' Welsh dry steam coal with pitch and carbonizing the ovoid so produced.

### *Opencast Working and Drift Mining*

Where a seam of coal lies or extends near to the surface it may be reached by surface working, known as opencast working, instead of by shafts and underground working. In this process the top soil and subsoil are removed and stacked separately so that they may be replaced later when the process is complete. The 'overburden', consisting of the material lying between the subsoil and the coal, is then removed and the coal dug out with mechanical shovels. The overburden is then replaced, followed by the subsoil and top soil. Overburden may be removed to a depth of 150 feet or more, and the operation is usually carried out on a trenching system so that one trench is filled as the next is dug.

Opencast working was developed during the war in order to supplement supplies of deep-mined coal. It is an expensive process, only justified when the supply of coal, and not the cost of producing it, is the first consideration. In 1945-46 nearly 8 million tons of coal were obtained by this means and sold at a loss of about £7 million (17*s*. 8*d*. loss per ton). In 1949-50, however, in consequence of improved methods and the increased selling price of coal, 13.5 million tons of coal were produced by opencast working and sold at a loss of just under £2 million (2*s*. 10*d*. loss per ton). Actual financial loss is not the only factor to be taken into account: the effect of opencast working on amenity and on agriculture must also be considered. First-class farmland may be destroyed in the process with resultant loss of produce and there is evidence that the restored land is less fertile than before. In consequence a difficult problem arises in determining at what point the gain in coal is offset by loss of agricultural produce, and by any permanent disfigurement of the landscape that may occur.

The quality of coal produced by opencast methods varies, but its nearness to the surface usually makes it unsuitable for specialized uses, such as coking and gas making. A disadvantage of this method of working is that the mechanical shovels tend to pick up a lot of clay and dirt with the coal, and coal preparation at opencast sites is usually impracticable. A simple system of picking and screening is resorted to, but good quality coal, such as anthracite, may be transported to colliery preparation plants for washing; this extra handling necessarily adds to production costs.

The National Coal Board do not undertake opencast working; it is done by contractors working for the Ministry of Fuel and Power. The Ministry prospect for suitable sites and, after consultation with the Ministry of Agriculture, the land is requisitioned under the powers of the Defence Regulations. Contracts are then placed with engineering firms on the basis of tenders for working the coal, usually at a fixed rate per ton of coal delivered. Compensation is paid to the land-owner for the land taken and to the National Coal Board for loss of profit and in lieu of royalties. The coal produced by opencast working is sold by the Board as agents for the Ministry. Price is determined not by the cost of producing the coal but according to the selling price of deep-mined coal of comparable quality; this accounts for the heavy losses incurred.

Shortage of coal supplies has caused the Government to postpone from time to time the termination of opencast working, and it is proposed to be brought to an end gradually after 1951. The House of Commons Select Committee on Estimates in the 1948-49 Session reviewed the production and marketing of opencast coal and recommended that Government policy should be reconsidered having regard to the disadvantages, including financial loss, which attend this method of coal production.

Drift mining is a method of getting coal by underground workings which are reached by a sloping tunnel instead of by shafts.

Chapter 5

# PLANNED PRODUCTION

### *The New Conditions*

The Reid Report made it clear that a general failure to plan adequately and far enough ahead had been one of the most obvious handicaps of the British coal industry. 'A vast programme of reconstruction of existing mines and the sinking of a number of new ones is now required, and the importance of avoiding the mistakes of the past, and of providing for the efficient lay-out of the underground workings and the orderly disposition of surface buildings should need no emphasis.' The Report recognized that this conception of long-term planning was directly opposed to a policy which required the earliest possible return on the capital expenditure incurred in reaching the seam, and emphasized that in planning more immediate objectives short-term expedients would have to be abandoned in favour of adhering to a set plan.

When the National Coal Board assumed control of the industry at the beginning of 1947 it was impossible to give immediate effect to these principles, admirable as they were. Before the Board could undertake large-scale reconstruction of the industry they had to concentrate upon short-term schemes which were capable of producing quickly the additional coal which the country desperately needed. The Board were faced with the fact that the industry's labour force had shrunk to an abnormally small number; that output per manshift had fallen steadily from 1936 to 1945; and that absenteeism, especially among faceworkers, had reached serious dimensions.

The only way to secure an immediate and substantial increase in output was to bring about a change of heart in management and men. The principles which the Board recognized should guide them as model employers in a new public service, are exemplified in their 1947 Annual Report:

> Machines, schemes and plans are nothing without the human genius, skill and effort which alone can put them to use. The human material of

an industry is more than its chief asset; it is the industry itself. It must be given opportunities for free and full development. Any organization must deal honestly with its workpeople if it is to prosper, but a nationalized industry, existing only to serve public ends, must set an example in the way it treats its employees, enlarges their opportunities and encourages their efforts.

In the coal industry, relations between employers and employed were not always good. Old grievances and old disputes had produced bitternesses. By meeting just claims, the Board hoped to infuse a new spirit into management and men, now partners in a new public service. For the mineworkers, this meant valuable concessions and an enhanced status. But with these privileges came new responsibilities. All in the industry must exercise self-discipline and moderation. In wages negotiations, for example, it is still necessary for the Unions to represent the interests of the men, but representations must be made with a due sense of public responsibility, since concessions for mineworkers can often only be obtained at the expense of the workers in other industries and indeed of the community at large. For management, too, the psychological environment has changed. There is a new pattern of loyalties, new responsibilities and new opportunities of service. All this is implied in public service—opportunities for all, sense of purpose, a responsible employer acting in concert with responsible employees in the joint service of the community.

The Board, acting in association with the Unions, sought to put these principles into practice by the means described in chapter 6 of this book. They adapted the industry's conciliation machinery to the new situation, provided facilities for consultation with the men at all levels of administration, and considerably increased financial provision for welfare arrangements, particularly for pithead baths and canteens. The principle of the five-day week was accepted, but the working of longer shifts, or an extra shift, had to be resorted to as a temporary expedient in order to produce more coal. Minimum rates of pay were increased and the wages of workers above the minimum were improved. The Board supplemented for mineworkers the benefits payable under the National Insurance scheme for industrial injuries.

In these ways, and at a cost of between £60 and £70 million a year, the Board sought to remove some of the bitterness and grievances of the past. The extent to which improvements in manpower and output in the first years of nationalization resulted from this policy cannot be determined, nor can it be said what the effect on the industry would have been if the Board had followed a contrary policy. The policy which the Board did follow represented a considerable investment in future goodwill, but the immediate results were disappointing.

Management no less than labour is faced with the need for considerable psychological adjustment to the new conditions obtaining in the nationalized industry. The changed character of labour relations, modern mining technique, and the work of reconstruction to be undertaken make new and exacting demands upon the limited numbers of qualified mining engineers available. There is an acute shortage of mining officials who possess the necessary technical qualifications and experience and who are able, at whatever level, to facilitate the smooth working of the new organization. This shortage will eventually be made good as a result of the training schemes and opportunities for promotion which are now available; in the meantime it impedes in many ways the refashioning of the industry.

### *Early Improvements*

One of the means by which increased output could be achieved relatively quickly was by mechanization at the coal face. Much modern equipment for mechanical cutting and loading is of American design and not always suited to the different conditions in British mines. In consequence it has been necessary to adapt existing or design new machinery suited to working conditions here. 'Meco-Moore' coal cutter-loaders which have been installed are now producing more than three million tons of coal a year, and at one colliery in the East Midlands Division the output of coal per manshift at the coal face has been raised to more than eight tons; the 1948 average for the whole country was 2.92 tons.

Increased output and reduced production costs have also resulted from the reorganization of underground transport. Haulage by pit ponies (sometimes by men) and endless rope systems is slow, of limited capacity, and wastes manpower. Replacement by mechanical conveyors and locomotive haulage operating in conjunction with high-capacity loading stations releases 'on cost' workers for upgrading to the face, and enables the increased output resulting from mechanization and other improvements to be handled efficiently at relatively low cost. During the first two years of nationalization the total length of conveyor belting installed in substitution for endless rope systems or to replace worn out belting increased from 9 to 13 million feet.

The introduction of locomotive haulage is less easily effected because many mines must be reconstructed before locomotives can be used in them. Most of the Divisions are occupied with schemes which include the driving of horizontal roads in the strata with a view to the use of mine cars of up to 4 tons capacity drawn by powerful locomotives; but many of these schemes cannot come into full operation for some years. There has, however, been a striking increase in the number of locomotives in use underground: by the end of 1948, 170 diesel and 15 electric battery locomotives were in use. Locomotive traction underground enables efficient man-riding facilities to be provided with a consequential saving in time and energy in travelling to and from the working face. In time the mechanization of underground transport will render obsolete the use of ponies and horses in pits.

Mechanization at the face and improved haulage systems have necessitated the enlargement and redesign of central loading points and of traffic movement at the top and bottom of the shafts in order to handle increased quantities of coal being produced. In order to concentrate working good haulage roads are being driven underground connecting two or more collieries and so enabling coal to be wound at one shaft instead of two or three as at present. In this way manpower can be saved and full advantage taken of skip winding and other up-to-date methods of raising coal. Reorganization of surface lay-out and the installation of coal preparation plant can then be planned on the basis of an adequate 'throughput'.

The increasing use of mechanized methods of coal getting, and the working of thinner and dirtier seams have made it essential to improve arrangements for coal preparation (see page 81). In recent years the installation of coal cleaning plant has failed to keep pace with the progress of mechanization. In 1930 less than a third of the coal produced was mechanically cut and only one-sixth was mechanically conveyed: a third was mechanically cleaned. By 1946 nearly three-quarters of the coal produced was mechanically cut and conveyed but less than half was mechanically cleaned. The installation of coal preparation plant stopped altogether during the war, and some plant that was available went out of use; other plant was inefficient in operation.

The Board's Coal Preparation Engineers surveyed the whole field of coal preparation and studied the technical problems involved. They overhauled existing plant, and by reorganization ensured that maximum use was made of such plant as was available. A considerable tonnage of coal is conveyed from collieries where cleaning plant is not available to collieries where it can be cleaned. The design of new equipment has been studied with a view to securing optimum efficiency, and wherever possible standard equipment has been introduced in order to speed up manufacture and installation.

Up to the end of 1948 twenty-six new plants costing about £5 million and having a capacity of 6 million tons a year had been started up. Additional modern plant of large capacity is on order or in course of installation and will be brought into use as the installations are completed. It will cost about £25 million and take many years to bring the Board's Coal Preparation Scheme into full operation, but when this is achieved 70 per cent of the output of coal will be cleaned mechanically, and the remaining 30 per cent will either be cleaned by hand or will not need cleaning.

*Long-term Projects*

During the first two years of nationalization the Board authorized expenditure amounting to £37 million on major capital schemes. The largest of these is at Nantgarw in South Wales, where nearly £5 million will be spent on reconstructing and completely modernizing the colliery, which was abandoned in 1927; in constructing coke ovens, which will manufacture blast furnace coke from the high quality coal produced in the colliery; and in installing by-product plant which will produce considerable quantities of gas, tar, refined benzole and sulphate of ammonia

Interesting examples of large-scale reconstruction and re-organization are at two collieries in Lancashire. At Bold colliery the sinking of a new shaft, the introduction of horizon mining, locomotive haulage, and skip winding will cost about £4.6 million and increase the output of the colliery five-fold, to more than a million tons a year. Similar reconstruction will be undertaken at Bradford colliery at a cost of £2.8 million and with a comparable increase in output. At each colliery coal will be delivered by conveyors direct from the coal preparation plant to nearby electricity generating stations operated by the British Electricity Authority. The coal from Bradford colliery will be conveyed by tunnel passing under the streets and buildings of Manchester. Expenditure exceeding £2 million has been sanctioned on a new pit at Rothes in Fifeshire and when completed this will have an estimated annual output of 1½ million tons. A million tons of coal a year is the estimated output from a new colliery in course of construction at Calverton, Nottingham, for which expenditure exceeding £1 million has been approved.

The Board had not only to increase immediate output and plan the re-development of the industry; they were faced with the need, hardly less urgent, to check the rising costs of production. Unification of the industry has made it possible to introduce a good deal of standardization of stores and equipment and so reduce purchasing costs and limit the variety and quantity of stores and spares which it is necessary to hold. Central workshops are being set up, and mobile repair gangs made available for dealing with major breakdowns. The adoption of standard voltages for electrical equipment has facilitated the progress of electrification, and a great deal of steam-raising and power plant which is obsolete or uneconomic is in course of being replaced or overhauled. Modernization of the power systems of the coal industry will save both fuel and labour, and effect an appreciable reduction in the cost of producing coal. At a single colliery in the West Midlands Division the installation of new water-tube boilers is estimated to have reduced fuel consumption by more than 300 tons a week. At a Scottish colliery, surface electrification has reduced the cost of producing coal there by 3*s.* 1*d.* a ton.

Further economies could have been made in the first years of nationalization if the Board had been free to close down pits which operated at a loss. They were unable to do so because the demand for coal was so great that it had to be met even by uneconomic working. In any case it would have been contrary to the policy of a nationalized undertaking to have created social and economic upheaval by the widespread closing of pits. In general the Board only closed pits when they were satisfied that the bulk of the labour displaced could be used more productively elsewhere. During the first two years of nationalization over sixty pits were closed because their working was unprofitable to an extent that did not justify their continuance notwithstanding the need for coal. This was usually in consequence of the coal seams having been worked out. Proposals to close pits were discussed with the men affected and they usually recognized the necessity of closing the pits and acquiesced in the decisions taken.

One of the Scottish coalfields affords a striking example of the difficulties associated with the closing of uneconomic collieries. In the Central area of Scotland there are pits in which seams are being worked 17 inches to 24 inches thick, the workings are wet, and surface conditions, especially housing, are extremely bad. Improvements are economically impracticable because the seams are almost worked out. If it had been possible to close 19 of these collieries in 1947 there would have been a saving of over £500,000 (equal to 9*s.* a ton) and the Divisional profit would have been increased by 35 per cent. But closure would have meant the displacement of 6,000 miners and their families; many of the men could not have been reabsorbed immediately at other Scottish pits, and their skill and experience would probably have been lost to the industry. Closing the pits would also have meant a loss of 5,000 tons of coal a day, and this coal was needed however uneconomically it was produced. The Board decided to continue working these unprofitable pits for the time being and to encourage the migration of labour to Scottish coalfields in which large-scale development is justified by coal reserves and the opportunity for planned development. The task is an enormous one involving the movement of 100,000 people over a period of ten years. In addition to sinking new pits in the developing areas it will be necessary for new towns to be built in order to provide for the well-being of the men, women and children affected by such a large-scale movement of population.

*The National Plan*

Major reconstruction of collieries and the sinking of new pits necessitates considerable capital expenditure and the diversion to long-term projects of labour and materials which the country cannot easily spare. It is therefore essential to resort to planned development so as to ensure that the country's present limited resources are not overtaxed and that the future pattern of the industry matches the nation's over-all requirements. In long-term plans the geological distribution of the nation's coal reserves must obviously be a governing factor, but it is important to determine not only where coal is to be found, but where coal of the right kind can be produced in the right quantities and at the right price. This in turn necessitates consideration of the economics of transporting coal by rail, sea and road, and of the practicability of siting collieries near to markets so as to reduce transport costs to a minimum. Future demand for coal and for particular types of coal must be estimated, and this cannot be done without due regard to the price at which coal can be sold and to competition from alternative sources of fuel and power.

Much is already known from the work of the Government Geological Survey, and from other sources, as to the nation's coal reserves, but the information available needs to be augmented, and in some cases to be checked before it can be used as the basis of forward planning. To this end the National Coal Board are carrying out an extensive programme of boring, on a scale three times greater than before nationalization took effect. In compiling records of the distribution and physical and chemical properties of the coal reserves the Board will have the assistance of the Geological Survey in the interpretation of data and in the exploration of some of the unproved areas. The programme of boring must itself be carefully planned in order to derive maximum benefit from the necessarily slow and costly processes involved.

This planning is facilitated by modern geophysical methods which make use of scientific apparatus in order to gain approximate knowledge of geological structure before boring is started.

Having determined where the coal is, and its nature, the extremely difficult problem arises of estimating the inter-related factors of rate of output and cost of production. This at once raises a whole series of questions which do not admit of any one answer. There will not always be a seller's market for coal and the nature and extent of consumer's demands must be taken into account in order to ensure that output generally and in particular coalfields is related to the demands of consumers at home and overseas. Under-production would prejudice the industrial and domestic well-being of the nation, and over-production would adversely affect the well-being of the industry, particularly in view of the practical limitations upon stocking coal. Planning necessitates study of the varying costs of producing coal and of the productive efficiency of the different coalfields. Decisions have to be made as to the extent to which some collieries may be run at a loss in order that the necessary volume of output may be maintained.

Since wages form so large a part of the cost of producing coal any scheme of development must be based upon assumptions as to future wages policy. These assumptions are particularly difficult to make since the whole wages structure of the industry needs completely revising, and the final pattern will depend upon the outcome of negotiations with the trade unions. One feature of labour costs which must be taken into account is the existing trend towards equalization of wages throughout the industry. The trend is towards a levelling up from the lower to the higher rates of pay and this must inevitably increase production costs if it continues. It by no means follows that a contented labour force (and therefore greater production) will result from wage equalization since this destroys differential rates of pay which are of long standing in the industry. Moreover, equalization would have the effect of checking the economically desirable migration of labour away from areas where wages are relatively low because costs of production are high to low-cost areas in which higher wages can be offered.

No small part of the price paid for coal by the consumer represents the cost of delivery to him. The problem of transportation must be considered in connection with any development scheme, not only because the advantages of production at low cost would be lost if heavy transport charges had to be met, but also because the capacity of the railways, and of other means of transport, to handle increased traffic from developing areas must be taken into account. It is clearly necessary to co-ordinate development plans for the railways and other basic industries with those for the coal industry. The new price structure for coal might be used to dissuade consumers from ordering coal from distant collieries when, perhaps with some adjustment of habit or equipment, equally effective use could be made of supplies nearer to hand. Industrialists may be induced to establish new factories in or near the developing coalfields by the prospect of relatively cheap coal supplies being available. The Board may decide to divide the country into a number of price zones and introduce a 'zone-delivered' price structure with a uniform delivered price within the zone for the same sort of coal. Future plans must have regard not only to the transport of coal but also to arrangements for its distribution and in particular to the adequacy of stocking facilities.

To some extent the future demand for coal is more easily assessed than such factors as output, and cost of production and distribution. It is known with reasonable certainty that the consumption of coal at gas and electricity works and in coke ovens is likely to increase, and that consumption by railways will diminish. As nationally devised plans for the future of those industries take shape the nature and extent of their demand for coal will be known. Consumption for general industrial and domestic purposes can also be assessed with a fair degree of accuracy, but the future demand of the overseas market is much more difficult to measure. Overseas buyers are concerned only with quality and price and can turn to alternative sources of supply if they are not satisfied. There is likely always to be an overseas market for the high-quality steam and coking coals produced in this country, but the need to maintain exports will make it imperative to find overseas markets for general purpose coals, and this will have to be done in the face of keen competition from other countries supplying coal. So far at any rate as the Board are concerned little influence can be exerted upon the overseas market except by supplying the right coal at the right price. Home demand can be influenced by a price system which matches price with quality, attracts demand to coals which can be economically produced and transported, and which discourages the use of high-grade coal which is in short supply because of limited reserves or the demands of the export market.

These factors of output and cost, demand and price, are fundamental in planning the development of the coal industry, but they cannot be divorced from wider considerations linking the coal industry with the economic life of the whole nation. Inordinate demands must not, therefore, be made upon labour and industrial resources so as to divert a disproportionate part of the nation's wealth to schemes of capital development. Whatever plans are made cannot disregard the social and economic consequences of abandoning old coalfields and exploiting new ones. The replanning and distribution of the collieries must also be undertaken with due regard to the requirements of national defence.

From an appraisement of these diverse factors a general scheme of development emerges as the national plan for the coal industry, on the preparation of which the National Coal Board have been actively engaged. The Board's Director General of Production has described the plan as 'allocating production between the various coalfields, determining the price structure, settling the extent of reconstruction work, allotting the capital finance required, and where necessary redistributing labour'; and constituting a co-ordinated scheme for giving effect to these purposes. In order to ensure a uniform basis of estimation the plan will relate to the circumstances which it is assumed will exist about the year 1960. The plan must, however, be flexible, for so many of the factors to be taken into account are themselves changing and their interaction must be frequently re-examined. When all the estimations have been made and decisions so important to the nation's well-being have been taken it will remain to implement the plan. Of this the Director General of Production has said:

> Without hard work the plan will be so much waste paper. From engineers, flexibility of mind and receptiveness of new ideas will be needed; improvements in standards of mining and other techniques must be achieved. In the management of the collieries there must be discipline since only with discipline can come efficient organization. But above all let us acknowledge that no amount of reconstruction will be of avail unless it is backed by hard work; hard thinking work and hard manual work.

*Science and Mining*

The coal industry had not, prior to nationalization, made any considerable provision for research and scientific control, though some work had been undertaken by the British Colliery Owners' Research Association (B.C.O.R.A.) under the direction of the Mining Association, and a few of the larger colliery concerns had laboratory services. The British Coal Utilization Research Association (B.C.U.R.A.), which represented many interests connected with the use as well as the production of coal, also carried on important research activities. Nationalization afforded the opportunity for a comprehensive scientific service to be made available for the industry as a whole, including the associated carbonization and by-product activities. The 'vesting' provisions of the Nationalization Act transferred some scientific equipment to the Board, and the property and staff of B.C.O.R.A., were taken over by arrangement with the Mining Association. This formed the nucleus of the Board's scientific organization.

A Central Research Establishment has been opened by the Board near Cheltenham and operates under the control of Headquarters and in close association with the Divisions. At this establishment fundamental investigations are undertaken of problems associated with mining and the treatment of coal products. In addition technical work is undertaken in connection with mining processes, coal preparation, carbonization, by-products, and the development of scientific instruments. Research of particular importance is concerned with the blending of coals which are in short supply, and scientific work on the technique of manufacturing carbonized briquettes will result in the economic use of fine coal. Greater efficiency in mining techniques may be expected to result from the methods of operational research which are being applied in the course of investigations at collieries by mobile scientific staff.

The Board's Scientific Department supervises and co-ordinates the research and general scientific work undertaken within the Board's own organization, and also acts in close association with B.C.U.R.A., the British Coke Research Association, and the Coal Tar Association. Close contact is maintained with the Medical Research Council and with the universities and hospitals, particularly in regard to work on those industrial diseases which occur in the coal industry. The Scientific Department is represented on the Scientific Advisory Council which is established by the Minister of Fuel and Power to co-ordinate research into problems affecting all kinds of fuel.

In the Divisions one of the principal applications of scientific method is to the control of mining conditions, and valuable work has already been done on new methods and instruments for the detection and measurement of mine-dust in roadways and fire-damp in the pits. Over a million samples of mine dust are analysed each year, and instruments are being designed which will enable rapid assessments to be made in the mine of gas concentrations, so that swift action may be taken where conditions are unsatisfactory. Apparatus has also been developed for accurately testing the efficiency of miners' lamps, and important work has been carried out on the control of underground heating in the seams.

The Divisional scientific service also provides quality control by sampling and testing the coal produced. Prior to nationalization about a quarter only of the coal produced was regularly sampled and analysed, and the Board plan to make adequate arrangements for effective quality control at each colliery. Until this comprehensive service is available, representative samples of all the marketed grades of coal are tested regularly. More than 6,000 named coals have been chemically surveyed in order that their qualities may be taken into account in the new price structure (see page 139), which will relate price to quality. Quality control also extends to the effective design and use of coal preparation plant, and preliminary scientific analysis is of considerable help in determining the type of coal preparation plant most suitable for the coal produced at a particular colliery. Already considerable saving has been effected in the amount of coal discarded in the operation of cleaning processes.

The Divisional scientists also undertake control of the quality of goods purchased by the Board and make studies of consumption rates. Scientific methods have been of great value in estimating the efficiency of colliery boiler plant. The incidence of mining diseases, methods of recording accidents, and the prevention of pollution by mining and other effluents are other matters in which scientific investigation is proving to be of value.

The Board have taken over from the Department of Scientific and Industrial Research the organization known as the Coal Survey, and the cost of this Survey is now met from the revenues of the Board instead of from Treasury funds. The work of the Coal Survey is closely linked with the work of the Government Geological Survey, and is more especially concerned with analysing the samples of coal produced in the course of boring operations and the sinking of pits. The knowledge so gained by scientists of the structure of the seams and of the quality of the coal in them is of the greatest importance in planning new developments.

Chapter 6

# THE BOARD'S EMPLOYEES

## *Manpower and Training*

Mechanized coal production cannot succeed unless men are available in sufficient numbers to work the machines with skill and diligence. Over the last quarter of a century the numbers of mineworkers have steadily declined, and the drift of men away from the coal industry has become a trend most difficult to check. In the year 1920 there were 1,226,900 workers employed above and below ground in coal production; by the end of 1946 this number had shrunk to less than 700,000. Unfortunately it was the younger and more active men whom the industry could not attract or retain. In 1931 only 34½ per cent of those employed were over 40 years of age, but by 1945 the over-40s had increased to 43½ per cent of the total, and the industry was forced to rely to an increasing extent upon men who had passed their prime in physical energy.

There were many reasons why this decline set in. The miner's work was hard and dangerous, but his pay ranked low in the scale of industrial earnings and his chances of unemployment were above average. The development of means of transport and the diversification of industry encouraged men to drift away from the isolated mining communities, and boys were discouraged from entering the pits. During the years 1926-28 there had been an abnormally low birth rate in the mining districts and this was followed, fourteen years later, by an abnormally low rate of juvenile entry to the industry.

Manpower difficulties were sharply increased in the period 1939-41 when numbers of men left coal-mining to join the Forces or take up better paid work in the war industries, and the labour force fell by more than 10 per cent during this period. In consequence an Essential Work Order of 1941 (superseding an Order of 1940 which did not have the desired effect) prohibited men from leaving the coal-mining industry without Government consent. Efforts were made to recall miners, men were directed to the industry, and 'Bevin' boys worked in the pits instead of serving in the Forces. Notwithstanding some success from these measures the number of mineworkers continued to decline. The Essential Work Order was revoked in 1946, and after that date the industry was protected by a 'ring fence scheme' imposed by a Control of Engagement Order. The effect of this was that persons normally employed in the coal-mining industry could only be engaged for employment outside the industry with the consent of the Minister of Labour. This restriction was brought to an end on December 31, 1949.

Apart from securing a steady flow of recruits to the industry to replace wastage, it is important to increase the number of face-workers, and more especially the proportion of faceworkers to other mineworkers. It is the faceworkers who get the coal, and the cost per ton at the pithead will be relatively lower when the numbers of men engaged on haulage and other ancillary duties are small in proportion to the number of faceworkers, and when out-of-date systems which are wasteful of manpower are replaced by modern systems of working. During the first two years of nationalization the number of faceworkers steadily increased, and at the end of 1948 the 296,000 faceworkers represented 40.77 per cent of the total number of wage earners on colliery books.

On the nationalization of the coal industry in January 1947, the number of mineworkers was the lowest on record for nearly half a century, and wastage in the previous year had exceeded recruitment by more than 4,000. The Government hoped to have 730,000 men on the colliery books by the end of 1947. After allowing for wastage (estimated at more than 60,000 men) this could only have been secured by recruiting 100,000 men and boys. The hoped for recruitment was not achieved and wastage was more than expected, but on balance the year showed a net gain of 26,165 men and boys. This was a promising start, but it included over 41,000 ex-miners recruited from the Forces and from other industries—sources that would diminish rapidly in subsequent years. More than 6,000 Poles were also recruited, and though there were some objections to their employment, they were usefully absorbed into the industry. A welcome feature of 1947 was a small increase, as compared with 1946, in the number of juveniles recruited notwithstanding the raising of the school-leaving age in that year. The manpower target for 1948 was set by the Government at 750,000, but the sources of manpower were drying up and notwithstanding a vigorous recruiting campaign and the engagement of foreign workers it was not possible to raise the number of men on the colliery books at the end of 1948 beyond 726,300.

Recruiting mineworkers is not just a matter of finding the men and passing them into the pits. They have to be trained, and when trained the places where they can be employed are limited by the men's experience, by the numbers of men which the pits can absorb, and sometimes by objections to the employment of foreign labour. Housing shortage is another limiting factor, and in the effort to overcome this difficulty 25,700 mineworkers were accommodated in Government hostels in 1948 at a cost to the Board of £900,000.

The increased use of machinery in the mines has emphasized the importance of preliminary and specific training for inexperienced mineworkers in order to prevent accidents and ensure efficient use of the machinery provided. Specialized training and educational facilities are also necessary to ensure that tradesmen, supervisors, managers, and engineers are available in sufficient numbers and with appropriate technical and other qualifications.

The training of juvenile recruits to the mining industry was examined by the Forster Committee, which reported in 1942, and in doing so recognized that a greater sense of economic security was essential in order to attract youths to the industry. The Committee recommended an eight-week course of general instruction and practical demonstration for juvenile recruits, together with payment of travelling expenses and the provision of hostel accommodation where necessary. The Reid Report stressed the fact that proposals for the future of the industry would not be complete without adequate training provision for the more exacting needs which new methods of work and a higher standard of mining practice would require.

Some of the larger collieries had operated training schemes for a number of years, but they were the exception and at the date of nationalization many comprehensive training schemes drawn up by the colliery concerns were no more than paper projects. Before training could be commenced on any substantial scale it was necessary to train the personnel who were themselves to act as trainers and supervisors of training at collieries. For this purpose the Board took over from the Ministry of Fuel and Power a Staff College which had been established in 1946 at Nuneaton, and (until September 1949) provided courses there for Training Officers, Training Centre Managers, Recruitment Officers and Welfare Officers. Pre-entry training was also necessary for the foreign workers recruited. Most of them knew no English and some were illiterate in their own languages; in consequence they were unable to assimilate preliminary training and would have been a danger underground. Before proceeding to adult training courses they went through courses of from seven to ten weeks pre-entry training at educational centres which the Board established. Remarkable results were achieved in teaching the English language and way of life to nineteen different language groups in conjunction with physical training and instruction in manual work.

Training Regulations made in 1945 contain a general requirement that no inexperienced person may be employed in or about a mine, except under competent instruction and supervision, unless and until he has been adequately trained and is competent to work without supervision. No one without previous underground experience of mining may be employed on work below ground in a coal-mine unless he has completed a prescribed course of training in practical mining work and theoretic instruction. The minimum period for such training is normally 264 hours, but as a temporary expedient the minimum period of training for adults has been reduced to 112 hours. Work at the coal face may not be undertaken by anyone without previous experience of such work until he has been employed for at least 80 days on work below ground in a coal-mine, at least 60 days having been spent at a training face under the close personal supervision of an instructor or supervising workman. (In certain circumstances part of a production face may be set aside for training purposes.) The Training Regulations may be relaxed in case of emergency, and the Minister has power to vary their requirements in particular cases.

Preliminary training is given by the Board in over 70 Group Training Centres, and these are supervised by Training Centre Managers and provided with a complement of training staff. Three of the centres are residential centres for juveniles. Special underground galleries provide facilities for training in the various mining processes, and at some centres coal faces are available at which coal getting, shotfiring and conveyor practice are demonstrated. Recruits who have reached a satisfactory standard of work at the end of the minimum period of coal-face training are transferred to productive work, but if they are employed in getting coal they must work for at least two years (less the period spent in training) under the supervision of an experienced coal-face worker before being permitted to work alone or to be in charge of any other persons employed in getting coal. The preliminary training given to juveniles considerably exceeds the minimum requirements of the Training Regulations and is closely integrated with class work provided by the local education authorities.

Training must also be provided for the mechanics and electricians needed to service colliery machinery, and for the men who operate coal-cutting machines and power loaders. This is undertaken in courses varying from 2 to 26 weeks in length at a Mechanization Training Centre at Sheffield which the Government set up in 1943 and which the Board have taken over. The Board have also introduced an apprenticeship scheme in order to meet the demand for skilled tradesmen. Boys of 15 or 16 years of age work as apprentices and supplement this training by part-time instruction at Technical Colleges before becoming improvers and then qualified skilled tradesmen.

The Board's training arrangements are backed by educational schemes designed to promote the efficiency of employees whatever their work may be, and to make it possible for youths of outstanding ability to fit themselves for the highest appointments in the industry. Scholarships provided by the Board, covering living costs, tuition fees and vacation visits to collieries, are available to boys leaving secondary schools, as well as to men and boys employed in the industry, to enable them to take mining and mechanical engineering degrees at the Universities. After graduation, directed practical training is provided in the industry and employment offering first-class prospects is available for mining surveyors and mining, mechanical and electrical engineers. Arrangements are made for exchanges of students between this country and Continental mining centres.

Employees are encouraged to attend full and part-time courses in technical and managerial subjects, whether provided by the Board or by Technical Colleges and Universities. Courses are arranged in the various Divisions at which problems of labour relations are studied. At annual summer schools held at the ancient Universities, miners, technicians and administrators meet together to study the problems of the nationalized mining industry.

In 1938 the Royal Commission on Safety in Coal Mines recommended that responsible officials, tradesmen and technicians should hold statutory qualifications of a higher standard than the deputy's certificate. Owing to the outbreak of war this recommendation was not followed up, but the Coal Industry Act 1949 has enabled the Minister of Fuel and Power to prescribe statutory qualifications to be held by any person employed in the coal industry. The Board plan to recruit, educate and train 4,000 men and boys a year as certificated electricians, mechanics, overmen, deputies, and shotfirers, and about 200 a year as colliery technicians, surveyors and undermanagers. The education and training will normally last five years or more and will combine theoretic and practical training in all the fundamental techniques of the mine.

*Output and Absenteeism*

Output per manshift (usually abbreviated to OMS) is the yardstick of the coal industry, and measures for a given period of time the amount of coal produced in one shift by one worker. Reference has already been made (page 34) to the extent to which the increase in OMS in Great Britain during the ten years ended 1936 was surpassed in foreign coalfields of comparable working conditions, notably the Dutch.

1936 was a peak year for OMS, not only in this country but in most of the other coal producing countries as well. In Great Britain in 1936 OMS for all workers was 1.18 tons, but by 1945 had declined to 1.00 tons. Subsequently there has been slow but steady recovery to 1.16 tons in 1949. OMS for faceworkers was 3.02 tons in 1949. The decline in OMS during a period of about ten years from 1936 took place notwithstanding steady increase in mechanization in the mines. Factors contributing to the decline have been the drift of the younger and more active men from the industry and the fact that the seams are becoming increasingly difficult to work.

It is important to note that OMS takes account only of shifts worked, and not of time lost through absenteeism, holidays, disputes, breakdowns, etc. Losses of output due to these and other causes are reflected in the annual output of coal per wage earner employed in the industry. In 1939 the output of saleable coal per employee amounted to 302 tons, and by 1946 this figure had declined to 260 tons. There has since been some recovery and output per employee amounted to 282 tons in 1949.

The greatest single cause of loss of output is absenteeism, this being the failure of a mineworker to attend a working shift. If the failure to attend is for a good reason, such as illness, absenteeism is called 'involuntary'; otherwise it is 'voluntary'. Voluntary absenteeism is one of the most serious and intractable problems which faces the mining industry. Where faceworkers are concerned it is particularly serious, for their absence not only means the loss of their own output, but disorganization results in the general working of the pit. Modern methods of mining depend upon strict observance of a continuous cycle of operations, and a high percentage of absenteeism may result in loss of production over a whole 24 hours cycle of operations (see page 73).

During recent years the average annual percentages of absenteeism have been as follows:

| | 1943 | 1946 | 1947 | 1948 | 1949 |
|---|---|---|---|---|---|
| Faceworkers | 15·1 | 19·2 | 15·0 | 14·1 | 14·94 |
| All workers | 12·4 | 16·0 | 12·4 | 11·5 | 12·34 |

These figures show an improvement in attendance since nationalization, but the extent of absenteeism throughout the industry is a matter for very serious concern. If the extent of absenteeism among faceworkers could be reduced by one-fifth the result would be an additional annual output of coal amounting to approximately seven million tons.

Many reasons have been advanced to account for absenteeism among mineworkers. One is that the spending habits of the miners have not kept pace with wage increases. In consequence some are content to absent themselves from one or more shifts in a week and still earn enough to maintain the low standard of living to which they became accustomed by economic need in the past. Housing conditions in mining villages are usually such as to offer little encouragement to the adoption of a higher standard of living. Again, the employment of miners' wives and daughters in local factories may result in an adequate family income being available without the need for the miner to work full shifts. In either case he is using his increased wages to purchase leisure rather than goods. To some extent full working makes new demands upon the miners, for during the inter-war years he was often prevented, by unemployment or short working, from working the full number of shifts even if he was willing to do so.

The fact that there are no breaks now caused by unemployment or short working time, and that the men are older on an average and therefore less able to bear the strain of working, is not as considerable a factor in absenteeism as might at first be thought. It is not the fact that the workers most addicted to absenteeism are old and tired, or lazy. There is more absenteeism among the young than the old miners. These young miners often work hard and well on piecework for three shifts or so in a week and then absent themselves, having earned as much as they need for a week's spending. Another factor in absenteeism is that mining is a hard and dangerous occupation, often carried out under conditions of considerable discomfort, and if a man does not feel at his best his disinclination to go down the pit may be reinforced by the thought that he will be a danger to himself and others if he does so when he is not a hundred per cent fit. Moreover, a miner loses the whole shift if he arrives at the pithead after the time for winding men into the pit has passed, and the very early hour of starting work on the main coal filling shift is conducive to losing time.

The problem of absenteeism is difficult of solution. Bonus payments for good attendance have not resulted in conspicuous success; increased pay has not provided a stimulus to work full time; and any decrease in wages or conditions of service could not be justified and would be strenuously resisted, and would in any case tend to reduce the manpower available. Absenteeism is almost invariably spoken of in connection with mining, but in fact it is by no means peculiar to the mining industry; it happens that the incidence and the consequences of absenteeism by mine-workers are measurable from day to day with a fair degree of accuracy and readily demonstrated. Absenteeism is one of a number of modern industrial problems as baffling as they are serious. It is, in the main, a human problem and should be looked upon as a measure of management efficiency. When success has been achieved in smoothing out the general economic and industrial maladjustments of the post-war world it may well be found that the particular difficulty of absenteeism in the mining industry has been solved in the same process.

*Wages and Conditions of Services*

An account has already been given in Chapter 1 of some of the agreements negotiated in the past which have resulted in the present extremely complex structure of the wages system in the coal-mining industry. Briefly, and with some sacrifice of accuracy in favour of simplicity, the system is as follows.

Mineworkers are paid piecework rates (i.e. according to the amount of coal gotten or work done), day-wage (i.e. shift) rates, or weekly rates. In very general terms it is true to say that face-workers are paid piecework rates; packers, rippers and underground repairers are paid piecework or day-wage rates according to local custom; other workers, such as the men on haulage and most surface workers (other than foremen), are paid day-wage rates. Deputies are usually paid by the day but overmen are paid by the week.

Piecework rates may be based on the output of an individual worker or of a team of men. Where piecework rates apply they are negotiated between management and men, and vary according to the nature of the seam to be worked, the difficulty of working, and the bargaining skill of the parties. The rates may be based upon the tonnage of coal or yardage of material cut, or upon the number of yards of roof ripped or floor cut, and extra payment may be made for exceptional working conditions, such as defective roof or wet workings. Piecework rates may be limited to a particular pit or seam, but day-wage rates (which are also settled by negotiation) normally apply to a whole district. Day-wage rates may be supplemented by additional payments in particular cases, e.g. for filling off a face.

Until 1944 a miner's wage was made up of a basis wage (i.e. his day-wage or piece rate), a percentage addition to the basis wage, the amount of which depended on periodical district ascertainments, and various flat-rate additions and allowances. One of the flat rates, viz. the war addition, was related to the cost-of-living index figure. An account has already been given in Chapter 1 of the basis wage and of the system of wages ascertainments by which district percentages are calculated.

In 1942 a Board under the chairmanship of Lord Greene established the first national minimum weekly wage for adult mineworkers, and recommended that the district percentage additions should not be allowed to fall below the level which they had then reached. The National Reference Tribunal under the chairmanship of Lord Porter introduced in 1943 a national minimum weekly wage for youths, and in 1944 the Tribunal raised the national minimum weekly wage for both adults and youths. In making these awards the Tribunal stressed the need for a thorough overhaul of the wages structure of the industry. A guaranteed wage, whereby a workman did not lose wages if for some reason work was not available, was introduced by the

Essential Work (Coalmining Industry) Order, 1941, and was continued with modifications, by later Orders. When the Essential Work Order was revoked the guaranteed wage provisions then existing were continued by a decision of the Joint National Negotiating Committee. The second Porter award, made in 1943, granted one week's holiday a year and laid down a formula for determining the aggregate amount of holiday payments to be made. The method of distribution of the aggregate amount was agreed by the Joint National Negotiating Committee.

In 1944 wages ascertainments agreements were suspended, and percentage additions then payable were merged in day-wages or piece rates. Most flat-rate additions, other than the war addition, were also incorporated in wages. Skilled craftsmen and certain other classes of adult workmen (varying according to district) were granted an increase of 1*s.* a shift. The agreement provided that so long as it was in force changes in wage rates at a pit would not be sought, other than those normally made in respect of changed methods or conditions of working. The agreement was later modified to enable wage claims at pits to be resumed after the National Coal Board's pit conciliation machinery came into operation.

The Board agreed to adopt and be bound by the provisions of collective agreements existing on the primary vesting date, and by an agreement made with the N.U.M. in 1946 undertook to pay the men for six statutory and customary holidays a year in addition to the annual holiday for which the second Porter award had provided. The Board later increased the national minimum weekly wage to £5 15*s.* for adult underground workers and £5 for adult surface workers, with corresponding increases (varying according to age) for those under 21 years of age.

Statements of average earnings are not an accurate indication of the earnings of the 'average' miner, because they do not show the wide range of earnings between the highest paid workers and those who are on the minimum rates. They are, however, useful in showing changes which take place in earnings taking the industry as a whole. Where earnings per shift are given the weekly income obviously depends upon the number of normal and overtime shifts worked in a particular week. Average cash earnings and the value of allowances in kind in 1938, 1946 and 1948 were as follows:

| | *Per shift* | | | | | | *Per week* | | | | | |
|---|---|---|---|---|---|---|---|---|---|---|---|---|
| | 1938 | | 1946 | | 1948 | | 1938 | | 1946 | | 1948 | |
| | *s.* | *d.* | *s.* | *d.* | *s.* | *d.* | *s.* | *d.* | *s.* | *d.* | *s.* | *d.* |
| Cash earnings ... | 11 | 3 | 23 | 11 | 31 | 9 | 55 | 9 | 118 | 7 | 157 | 1 |
| Value of allowances in kind ... ... | | 5 | | 11 | 1 | 5 | 2 | 2 | 4 | 7 | 6 | 9 |

The allowances in kind are mostly coal which is supplied free or at reduced prices.

The amount of cash which the miner receives is affected by 'off-takes' for income tax, national insurance contributions (including contributions for additional injury benefit) and in some cases for trade union dues, rent (if a Board tenant), coal, and other items.

The five-day week agreement, which came into force on May 5, 1947, provided that the normal working week for underground workers should consist of five consecutive shifts of $7\frac{1}{2}$ hours each plus one winding time (i.e. the time taken to get a whole shift from the surface to the working level; over the country as a whole the time averages about 30 minutes). The actual amount of time spent at the coal face depends upon how long it takes to get the men from the pit bottom to the coal face and back again (see page 77).

In order to prevent loss of earnings the five-day week agreement provided that men who worked the full qualifying five shifts in a week should be paid a bonus. A man on daily wage rates receives a bonus equal to his average day-wage rate for the week; a man on piecework receives a bonus equal to 16 per cent of his earnings (excluding overtime) over the five qualifying shifts. It was agreed that full shifts should be worked in order to prevent loss of output due to a reduction in the normal weekly hours, and that by local agreement the miners' tasks should be increased in order to provide a fair day's work where existing tasks were insufficient; in general the response to this feature of the agreement has been disappointing. The agreement provided that the normal working week for surface workers should be $42\frac{1}{2}$ hours, exclusive of meal times, to be worked in five consecutive shifts of $8\frac{1}{2}$ hours.

Towards the end of 1947 the demand for increased coal production necessitated an extension of working hours. This was agreed to, without prejudice to the established principle of the five-day week, and it was left to local arrangement to decide whether to carry this out by Saturday working, or by means of an extra half-hour on the day from Mondays to Fridays, or by both methods. Provision was made for extended hours to be paid at the appropriate overtime rates. The maximum length of time during which the main categories of workmen may be below ground for the purposes of their work, or of going to and from their work, is limited by the Coal Mines Regulation Act, 1908 (as subsequently amended and applied) to $7\frac{1}{2}$ hours per day plus one winding time. In order to provide for the extended working-hours to which reference has just been made, an Order in Council having temporary effect has suspended the operation of the relevant sections of the 1908 Act.

The National Insurance (Industrial Injuries) Act, 1946, which replaced the former statutory provision for workmen's compensation, gave authority for Supplementary Schemes providing higher rates of benefit in particular industries, and the scheme for the mining industry was the first to be submitted to Parliament. The Scheme provides for a weekly supplement of 20*s*. to adults drawing injury benefit, and a weekly supplement to disablement benefit normally amounting to one-third of the National Insurance disablement benefit. A supplement, the maximum of which is 20*s*. a week, is paid to a colliery worker's widow who receives death benefit. These supplementary benefits are administered by the State as part of the general industrial injuries scheme, but they are paid for by contributions of 4*d*. a week from the men and 4*d*. a ton from the Board on the total saleable tonnage of deep-mined coal. In this way six-sevenths of the cost of the Supplementary Scheme is met by the Board. The Board also supplement Workmen's Compensation Act payments and have made special provision for ex-miners suffering from pneumoconiosis.

Men can become industrial casualties through redundancy as well as by reason of injury or disease. If redundancy of an exceptional nature results from reorganization and reconstruction of the coal industry the Board pay compensation to men displaced if they have been employed in the industry for periods amounting in all to not less than three years. Payments of 43*s*. 8*d*. for underground workers and 33*s*. 8*d*. for surface workers may be made for a maximum of twenty-six weeks of unemployment. These payments are in addition to national unemployment benefit.

*Trade Union Representation*

The Coal Industry Nationalization Act imposed upon the Board the duty of entering into discussions with organizations appearing to them to represent substantial proportions of their employees, or classes of their employees. The purpose of such discussions is the establishment and maintenance of joint machinery for 'conciliation' and 'consultation'. 'Conciliation' is the settlement by negotiation of terms and conditions of employment, with reference to arbitration in default of agreement. 'Consultation' is concerned with matters affecting the safety, health or welfare of the Board's employees, the organization and conduct of operations in which such employees are engaged, and other matters of mutual interest. The Board employ approximately 800,000 men and women in widely differing occupations and at various levels of income, and the establishment of 'conciliation' or 'consultative' machinery which is fairly representative of them without being hopelessly unwieldy is a task of peculiar difficulty.

The National Union of Mineworkers represent by far the greatest number of the Board's employees, and for this reason the N.U.M. play a leading role in negotiations and consultations between management and workers in the coal industry. The Deputies are represented by the National Association of Colliery Overmen, Deputies and Shotfirers (N.A.C.O.D.S.). This Association, as its title makes clear, also represents Overmen and Shotfirers, but as the N.U.M. also represent some of these men, the Board negotiate jointly with the N.U.M. and N.A.C.O.D.S., so far as Overmen and Shotfirers are concerned.

The majority of the managerial, professional and technical grades are represented by the British Association of Colliery Management (B.A.C.M.)—not to be confused with N.A.C.M., the National Association of Colliery Managers, which is a professional and technical association not concerned with terms of employment. Clerical and junior administrative staff are represented by the N.U.M., the Clerical and Administrative Workers' Union, and the National Association of Clerical and Supervisory Staffs. Among other unions with whom the Board negotiate in respect of technical and professional grades are the Association of Scientific Workers, the Association of Building Technicians, and the Association of Shipbuilding and Engineering Draughtsmen.

Considerable numbers of the Board's employees work in processes which are ancillary to mining, such as bricklaying. Their representation in negotiations with the Board is complicated by the fact that their wages and conditions of service are often regulated by agreements which are not limited to the mining industry. Methods of representation in negotiations with the Board are in process of being worked out in consultation with the N.U.M. and the other unions concerned. A National Joint Council, National Reference Tribunal, together with District and Plant Conciliation Schemes, have been established for workers at coke and by-product plants. Similar conciliation machinery is in process of being established for workers at briquetting plants.

*Conciliation*

When the Board took over the mines, conciliation machinery was already in existence, consisting of a Joint National Negotiating Committee (J.N.N.C.), a National Reference Tribunal, and a series of District Conciliation Schemes. These were adapted to the new organization, and comprehensive provision was made for conciliation machinery at the pits.

The J.N.N.C. as reconstituted on nationalization consisted on the employers' side of all the members of the Board (but with a quorum of three), and on the employees' side of fourteen members representing the N.U.M. (including representation of the Union's Area Executives). There was no direct Divisional representation on the Board's side of the J.N.N.C. Arrangements were, however, made for consultation with the Divisions on matters of national negotiation, and for attendance of their representatives at meetings when required. This arrangement was altered in 1949, when the Board's side of the J.N.N.C. was reconstituted so as to include representatives of the Divisions. The Board's side now consists of not more than fourteen members, of whom six must be members of the Board.

The J.N.N.C. are concerned with any matter relating to wages and conditions of service which is of national significance. Matters with which the J.N.N.C. have dealt include the application of the five-day week agreement, payment for statutory holidays, the extension of hours, increase in minimum wages, and compensation for redundancy The J.N.N.C. have usually been able to reach agreement on matters before them, but if agreement is not possible the issue in dispute is referred to the National Reference Tribunal. This Tribunal is appointed by the Master of the Rolls and consists of Lord Porter as President and two other members; its decisions are binding on both sides of the industry. No matter was referred to the Tribunal during the first year of nationalization, and only three matters were referred in 1948.

District Conciliation Schemes comprise a District Conciliation Board and a District Referee. The Boards are made up of representatives of the N.C.B. and the N.U.M., and they negotiate on matters affecting particular coalfields. Matters on which a District Conciliation Board fail to reach agreement (not being matters of national implication) are referred to the District Referee for decision. The scheme of District conciliation has been somewhat complicated by the fact that the Divisions of the N.C.B. and the Districts of the N.U.M. do not coincide in area, and a number of local adjustments have been necessary. The amount of conciliation work undertaken by the District Conciliation Boards has so far been small because district wages remain 'frozen', but the Boards have discussed the local application of national agreements, and have appointed Joint Dispute Committees and panels of Umpires in connection with the Pit Conciliation Schemes.

A uniform scheme for settling disputes at pit level is quite new. The Board and the N.U.M. have established a scheme by which disputes, other than those having district or national implications, are capable of being settled quickly by negotiation at the pit concerned, or failing agreement, by the award of an umpire. The procedure is well described in the following extract from the 1947 Annual Report of the Board:

> Any question in dispute at a pit must first of all be discussed as soon as possible between the workmen and the official immediately concerned. Failing agreement within three days, the matter must again be discussed between the men concerned and the manager of the pit or his representative. If agreement is still not reached, the men report the dispute to the local Trade Union official who decides whether the question shall be taken up by the Union. If further discussion with the management does not produce settlement, then the Trade Union official must write to the colliery manager asking for a meeting to be held between representatives of the Union and the management (or, if the question is one raised by the management, then the manager must write to the Union). This brings into operation the 'pit meeting' which must be held within five days of the request being made. At this meeting Area officials of the Union and Area or Divisional officials of the National Coal Board may be present. Agreed copies of minutes are given to both sides. If the dispute is still not settled within 14 days after the pit meeting it must then be referred to the joint secretaries of the District Conciliation Board, who must refer it immediately to a Joint Disputes Committee specially set up under the Pit Conciliation Scheme to deal with pit disputes. . . . If the Disputes Committee cannot agree on a matter put before them, then it is referred to an umpire selected from a panel appointed by the District Conciliation Board. An assessor nominated by each side must be present at all hearings by the umpire. This is the last stage in the Pit Scheme and unless it is decided that the dispute raises questions of principle applying to the coal field or country as a whole, the umpire's decision is final.

The extensive provision which has been made for the settlement of disputes between the Board and their employees should prevent strikes and 'go slow' working, but unfortunately this result has not always been achieved. The estimated tonnages of coal lost through disputes in recent years are as follows:

| | *No. of Stoppages* | *Loss of Output in tons* |
|---|---|---|
| 1946 | 1,329 | 770,000 |
| 1947 | 1,635 | 1,652,000 |
| 1948 | 1,528 | 1,062,300 |

Of the 1947 loss, 900,000 tons was accounted for by four strikes, of which the most serious was the strike at Grimethorpe Colliery arising out of the implementation of the five-day week agreement. Most of the 1948 strikes were caused by disputes about wages, but some were concerned with the closing of pits, and were therefore problems of consultation rather than of conciliation. Some improvement in the work of conciliation is becoming apparent as experience is gained in the use of the machinery available, much of which is new, but there is room for considerable further improvement.

*Consultation*

Effective consultation between management and men is one of the most likely means of gaining the good will and co-operation of the mineworkers. Men work better when they understand the significance of what they are doing and the general plan of what is intended, and they are less resistant to change if they understand and approve the need for alteration. Consultation also affords opportunities hitherto lacking for the men to contribute their own ideas and experience to the highly complex task of running a colliery.

During the war Pit Production Committees representative of management and men were set up under the Government's reorganization scheme for the coal industry with a view to promoting increased output. The National Coal Board have considerably developed arrangements for consultation by extending the system to all levels of administration and widening the range of subjects within the competence of the consultative machinery.

The basic unit of the new system is the Colliery Consultative Committee, the constitution of which is in accordance with a model prepared by the National Consultative Council. The colliery manager is always the chairman of the Colliery Consultative Committee, and he appoints three of its members, two of them being underground officials. Six members are elected by the mineworkers by secret ballot; two of them represent the faceworkers and the remaining four represent the other main grades of colliery workers. One deputy is elected by secret ballot of the deputies. The Board's mining agent, the N.U.M. Area Agent, and the Colliery Lodge Secretary are *ex officio* members of the Committee; others may attend on invitation if they have special knowledge of matters under discussion. Members of the Committee are elected for three years and one-third of their number retire each year. Workmen, other than deputies and managerial staff, who are members of the N.U.M. and have been employed at the colliery for not less than twelve consecutive months may nominate candidates for election to the Colliery Consultative Committee. The local lodge or branch of the N.U.M. select at least two of the nominees for each vacancy to be filled and a ballot is then held to decide which of the selected nominees shall become members of the Committee.

There are now Consultative Committees at most of the 1,000 collieries. In addition there are Area and Divisional Consultative Councils (with some local variations in structure) representative of the Board, the N.U.M., N.A.C.O.D.S. and the N.A.C.M., and these Councils consider at their respective levels the subjects appropriate to consultation. The National Consultative Council, of which the chairman of the National Coal Board is chairman, consists of six members appointed by the Board, nine appointed by the N.U.M., three appointed by N.A.C.O.D.S., and nine appointed by the N.A.C.M. The National Consultative Council studies the general working of the consultative machinery and has discussed such matters as output difficulties, improved mining methods, education and training, and the welfare, health and safety of mineworkers. A Committee on Safety and Health has been established and this Committee examines drafts of all new Safety Regulations. The Chief Inspector of Mines and representatives of the Minister of Fuel and Power attend meetings of the Committees when these drafts are being considered.

The Consultative Councils and the Colliery Consultative Committees may consider any matter relating to the operation of the collieries except wages and conditions of service: these are dealt with by the conciliation machinery. A subject particularly well suited to consultative procedure is colliery reorganization, especially when the closing of pits and the transfer of labour elsewhere is involved. The extent to which pits have been closed by the Board without serious trouble arising is an indication of the value of consultation. Other matters which have been examined in consultation are absenteeism, output, new mining methods, safety, the intake of juvenile labour, and training arrangements.

Consultation has not proved an unqualified success. It is a new technique demanding co-operation by management and men to an extent not hitherto customary in the mining industry, and in consequence it has been regarded with suspicion by some and with hostility or indifference by others. The fact that responsibility for managerial decisions rests with the Board has frustrated those who regarded consultation as a form of workers' control. Others have not appreciated the difference between consultation and conciliation, and have sought to use the Committees exclusively for the discussion of grievances. There was danger at one time of the Committees being regarded as an alternative to union machinery as a means of dealing with grievances, but conflict has been avoided by the association of the unions with the consultative machinery in the manner already described. Notwithstanding initial difficulties there has been growing appreciation of the part which consultation has to play in the industry, and the Board have sought by conferences and other means to improve understanding of the working of the consultative machinery.

*Welfare*

Prior to nationalization some of the large colliery concerns had established their own welfare schemes, but in general welfare activities were the concern of the Miners' Welfare Commission, the evolution of which has already been described (page 24). When the Nationalization Act was before Parliament it was decided not to bring the Commission to an end. Instead, it was provided that for the future the Commission should consist of a chairman and nine members appointed by the Minister of Fuel and Power without the former statutory requirement as to representation of various interests. The Minister was empowered to give directions of a general character to the Commission as to the exercise of their functions, and the Commission were empowered to act as the agent of the National Coal Board with respect to any matter relating to the health or welfare of persons in the Board's employment. The Commission were freed from the duty of appropriating funds to research into the health and safety of mineworkers. The consequence of these provisions was that two bodies—the Board and the Commission—were charged with welfare functions, for Parliament had specifically required the Board to have regard to the safety, health and welfare of their employees. There was a very large area of overlap, but the functions of the two bodies did not in all respects coincide. The Board's welfare duties extended to all their employees, whether or not they were mineworkers; the Commission's functions extended to mineworkers and their dependants.

The need to co-ordinate the activities of the Commission and the Board was obvious. This co-ordination has been provided for in a scheme approved by the Minister of Fuel and Power under which a joint authority—the National Miners' Welfare Joint Council—has been established, consisting of the Miners' Welfare Commission and two representatives of the Board. (The Board are, of course, already represented on the Commission.) Further integration has been secured by appointing a member of the Board (and latterly the Chairman of the Board) as Chairman of the Commission and of the Joint Council, and by transferring the Commission's staff (with a few exceptions) to the Board, who would give effect to the Council's decisions.

The function of the Joint Council is to act as the central authority for planning and directing welfare activities for the benefit of all workmen in or about coal-mines and their dependants, and others in the Board's employment. The Council are assisted by a Divisional Welfare Committee in each Division, these Committees being composed of representatives of the Board and of the Unions and others. They undertake the detailed planning of welfare activities in the coalfields, and in doing so they are assisted by Committees similarly constituted for the Areas. There are Local Committees responsible for the actual running of welfare schemes, such as recreation grounds, institutes, pithead baths and canteens. These Committees are also representative of management and men, and a close relationship is encouraged between them and the Colliery Consultative Committees.

The principal sources of income which the Miners' Welfare Commission bring to the Welfare Joint Council are two statutory levies, one of a penny on every ton of saleable coal produced and the other of a shilling on every £1 of (notional) royalty. These levies, together with income from investments, yield an income of just under £1.0 million a year. The Board have added an additional twopence a ton on saleable output of coal and this additional income, amounting to just over £1.5 million a year, will be devoted to capital expenditure on pithead baths, canteens and ancillary buildings. In addition the Board meet expenditure amounting to more than £2 million a year for the management and maintenance of pithead baths, the improvement of existing canteens, medical facilities, and other welfare services.

The only activities of the nature of welfare which are dealt with by the Board direct are housing, technical education, and health and safety research. The planning and co-ordination of other purposes connected with the education, social well-being, and recreation of the Board's employees are dealt with through the Welfare Joint Council. Welfare services so provided include pithead baths, canteens, recreation schemes, colliery institutes, community centres, rehabilitation centres, convalescent homes, and hospital schemes. Educational and cultural activities are encouraged, such as handicraft displays and musical and dramatic performances.

The best known of the welfare services are the pithead baths. There are now pithead baths available for nearly half-a-million mineworkers, and further development projected will entail capital investment exceeding £9 million by 1951. The Board have relieved the miners from contributing towards the cost of maintaining the baths, and the men have now to provide only soap and towels. During the war there was considerable expansion of the provision of colliery canteens, and now most collieries are provided with canteens, though some of them are on austerity lines. The Board have undertaken to meet the cost of capital improvements in raising such canteens to a higher standard, and employ Catering Officers who are available to assist the Committees in improving the service provided.

*Safety and Health*

Coal-mining inevitably entails a high risk of accidents, and in consequence the industry is a dangerous one. Approximately 600 men have been killed and 2,600 have been seriously injured on an average every year during the past ten years. There has, however, been steady improvement, particularly in the number of fatal accidents; the average number of men killed has fallen to 528 a year during the past five years. The principal causes of accidents are falls of roof, haulage mishaps, and explosions.

The control of undermined strata has been the subject of study for some years by the Safety in Mines Research Board. The staff of this organization has now been transferred to the National Coal Board, and research by the Board's mining engineers into the best methods of roof control is actively continued. Improved methods of stowing the waste from which coal has been excavated, and support at the face by hydraulically operated props and 'yielding' pillars, offer the best prospects of reducing the number of roof-falls and consequential accidents.

Explosions are caused by the ignition of inflammable gases given off by the seams, by the combustion of explosive concentrations of coal dust in the air, and by fires burning in the waste. Improved methods of underground ventilation, better methods of packing the waste, and energetic measures to suppress coal dust are reducing the risk of fires and explosions. Stricter observance of the Safety Regulations would reduce the risk appreciably, for it is distressing to note that many lives are lost through breach of the Regulations, particularly those which prohibit cigarettes and matches being taken into gassy pits. Modern haulage systems and the extended use of electric lighting underground will bring about conditions in which accidents of a general nature can be reduced to a minimum. The Board have appointed Safety Officers at Headquarters, Divisions and Areas, as well as for collieries and groups of collieries, and these officers are continually studying the causes of accidents, and devising and supervising the development of new safety measures. Rescue Stations are maintained in the coalfields at which trained personnel and equipment are available to be rushed to the scene of any accident which necessitates extensive rescue operations. Ambulance services are provided either by the Board direct or by arrangements with local health authorities.

Apart from danger of accident, miners run the risk of contracting certain diseases which are associated with mining. The most serious of these is pneumoconiosis, caused by inhaling coal or stone dust. Miner's nystagmus is a nervous disease which affects the eyes of some men who have to work underground in badly lit places. Dermatitis and infections of the knee, elbow and hand ('the beats') are other occupational diseases found among mineworkers. Research is being undertaken by the Board and other institutions into the causes, incidence and treatment of these and other diseases in mineworkers, and steps are being taken by mass X-ray, diagnosis and early treatment to detect and check disease in individual workers.

In view of the high accident rate and incidence of occupational disease the coal industry had not, prior to nationalization, provided colliery medical services on a scale in any way comparable with the standards of modern large-scale industrial concerns. The Miners' Welfare Commission had done useful work in establishing rehabilitation centres and convalescent homes, assisting hospitals and nursing associations, and providing or supplying ambulance, surgical and medical services, but there was nothing approaching a comprehensive mines medical service. To remedy this deficiency the Board have appointed a Chief Medical Officer and Divisional Medical Officers, and plan to appoint doctors for large collieries or groups of collieries. Medical centres with proper accommodation and the services of doctors and State Registered

Nurses are being provided and First Aid arrangements generally are in process of complete overhaul. Almost all new entrants to the industry now have pre-entry medical examinations (many of the examinations including a chest X-ray) to ensure that they are physically suited to mining work. Periodical examinations and the prompt treatment of injuries, even though slight, will reduce involuntary absenteeism through sickness and wastage of manpower through permanent disability, and will make mining more attractive to those who work in it and to those who contemplate doing so. For those who have suffered injury or disease residential centres and out-patient clinics provide modern methods of rehabilitation. Under approved conditions of work men permanently disabled by occupational disease or injury, including pneumoconiosis, paraplegia and other serious conditions, are being re-employed, and so are able to feel of value and service to the community in circumstances where formerly they would have been useless to themselves and others.

Chapter 7

# MARKETING

*Coal Classification*

In order to sell coal something must be known about its suitability for particular purposes. Coal is not homogeneous, and its properties differ from seam to seam and pit to pit. Whatever it is used for, the heat content (calorific value) of the coal is important, and the size of the coal must be suited to the consumers' plant or appliances. For some users other factors assume great importance: coal for gas making must provide a high yield of gas, and coal suitable for coke-ovens must produce a good quality coke sufficiently hard to withstand the burden of the blast furnace charge. For steam raising, heat content alone is most important, but if the coal is needed in large-sized lumps its tendency to break up must be considered if it is to be transported long distances or shipped abroad. Trade usage has made familiar such terms as 'coking', 'gas', 'household' and 'steam' as general descriptions of different kinds of coal, and the name of the colliery, and even of the seam from which the coal has been produced, has indicated to the trade its particular qualities. Coal salesmen have considerable knowledge and experience of the properties of different coals, and they provide a useful service in giving advice to the consumer on the suitability of certain coals to his plant. The trade nomenclature, however, is lacking in precision, and does not bring under a common heading different coals which would serve the same purpose equally well; it has resulted in unnecessary complexity with about 6,000 named coals on the market; and it does not take account of the fact that the quality of coal varies even in the same seam.

A scientific classification of coal is not easy because of the different nature of the properties which have to be taken into account. A useful basis for a classification is the volatile content of the coal—volatiles being that part of the coal substance which is driven off in the form of gas when coal is heated. If the volatile content is related to the 'caking' properties of the coal (i.e. its ability to 'cake' or form coke during the process of carbonization —see page 84), coal can be classified into a number of 'ranks'. This expression refers not to the quality of the coal in the ordinary sense of the word, but to the extent to which the primeval plant matter has been changed into coal. Coal of the highest rank is anthracite, which contains less than 10 per cent of volatile matter and is non-caking. Anthracite has a high calorific value, burns practically without flame, and is entirely free from smoke or soot. At about 15 per cent of volatile matter content, caking properties develop, and caking steam coal may have weakly, medium, or strongly caking properties. Next come the medium volatile coals with 20 to 30 per cent of volatile matter, and these include the best metallurgical coking coals. The high volatile coals contain more than 30 per cent volatile matter and may be either strongly, medium, weakly or non-caking. The best gas coals are included in this range.

This classification of coal by rank is by no means exhaustive, and other factors such as ash, sulphur, and moisture content have to be taken into account. Apart from the inherent qualities of the coal, the manner of preparation and size of the coal is important. In the past, terms indicating the sizes of the coal have varied throughout the country. The Board have adopted the Report on Size Grading made by the British Colliery Owners Research Association in July 1946, which recommended the following standard size groups and nomenclature:

| *Name* | *Screen Size* |
|---|---|
| Large Cobbles ... | 6″ × 3″ |
| Cobbles ... ... | 4″ × 2″ |
| Trebles ... ... | 3″ × 2″ |
| Doubles ... ... | 2″ × 1″ |
| Singles ... ... | 1″ × ½″ |
| Peas ... ... ... | ½″ × ¼″ |
| Grains ... ... ... | ¼″ × ⅛″ |

'Smalls' are coals which pass through a specified screen aperture but have no lower limit of size.

Having adopted standard methods of classification for the properties and sizes of coal, it is important to ensure consistent observance of these standards. Here the colliery laboratory is of service in providing regular analyses of samples and an effective quality control (see page 100).

*The Transport of Coal*

Coal is transported from the pithead to the consumer in railway wagons, motor lorries or canal barges, and is shipped from local ports. If the consuming point (e.g. coke-ovens and colliery brick-works) is very near the pit, conveyor belts may be used to transport the coal.

The bulk of the coal is moved by rail transport. At the collieries coal is loaded into railway wagons which are weighed and labelled and then worked through the colliery sidings to the main line system. Loads for large consumers, such as power stations or large coke-ovens, may be made up into single trains and taken direct to their destinations. Smaller 'block' loads are marshalled into trains and later split up for delivery.

There are several types of mineral wagons in service, suitable for different discharging arrangements. The modern standard wagon has five doors and can hold sixteen tons of coal. Other wagons may have only end doors, or side doors, or a combination of both. A popular type of wagon is the 'hopper' which is discharged through one or two bottom doors. Although there are not many in service, some wagons have no doors at all, but are designed to be emptied in rotary tipplers.

Some idea of the scale of inland coal movement is given by the fact that two-fifths of all railway freight is coal, and one half of the total number of railway wagons in service are suitable for coal traffic. The railways have a dual problem so far as the coal industry is concerned—they must keep the coal moving from the pits to the consumers, and they must keep the collieries supplied with empty wagons.

Before the war about half a million mineral wagons were privately owned, the remainder being owned by the railway companies. At the outbreak of war in 1939 the Minister of Transport took over control of the railways and requisitioned all privately-owned railway wagons in order to ensure that they were used to the best advantage in a national pool. When the railways were nationalized in 1948, the ownership of all these wagons (many of which had meanwhile vested in the Board) was transferred to the British Transport Commission. With certain exceptions, the use of privately-owned railway wagons on the nationalized railways is now prohibited. Now that both the coal industry and the railways are nationalized there are opportunities not previously available for developing a well-planned system of coal transport that will facilitate the smooth working and quick turn-round of mineral wagons. The use of larger wagons is desirable for more economic working of the railways, but one of the problems at many collieries is that the screens are too low to accommodate wagons of more than sixteen tons capacity. In addition, many retail coal merchants prefer small wagons which can be discharged quickly so as not to incur demurrage.

A considerable tonnage is moved in ships, either to destinations on the coast or abroad to foreign countries. Coal is usually taken by rail to loading appliances at the ports, but loading direct into ships is possible at colliery loading staithes. Whether the coal is intended for a coastwise destination or for export, the sale is usually made on a f.o.b. (free on board) basis, leaving the coal exporter or distributor to arrange shipment. It is their duty to charter and 'stem' the vessel (i.e. make arrangements for specified quantities and qualities of coal to be loaded into a particular ship). The vessel is then loaded either by tipping the wagons directly over the hold of the vessel or by conveyor belts; these belts are usually fitted with anti-breakage devices to prevent the coal from being broken up too much.

When the coal arrives at the consumer's yard or distributor's depot, it has to be unloaded. Ships are unloaded by cranes or grabs, and wagons may be emptied by small cranes or tipped. At a house coal depot, the coal may be shovelled directly into bags, which are placed on a lorry for distribution to the householder. During the summer-time some coal has to be put into stock. The various grades of coal have to be stored separately, and the stockpiles have to be arranged carefully, otherwise there is danger of spontaneous combustion. This summer stocking of coal is an important feature of the trade. It assists in maintaining regular production at the pits throughout the year, and it enables the merchants to meet peak demands in winter and to provide for any hold-up in supplies owing to fog or bad weather conditions.

Eventually the coal reaches the consumer. It comes to the large-scale consumer in shipload, trainload or truckload. It is brought to the domestic consumer in one- or two-hundredweight bags drawn in the familiar coal cart. To an increasing extent motor vehicles are being used, not only for retail delivery but also for bulk delivery.

*Coal Distribution*

On vesting date, the statutory selling schemes referred to on page 31 were dissolved and their staffs transferred to the Board. The assets and staff of wholesale and retail sales organizations which some colliery concerns maintained, were also transferred to the Board. Coal distribution was not nationalized, however, and in consequence the pre-vesting date practice continues, the Board undertaking coal distribution in common with many other wholesale and retail merchants.

The first sale of coal from collieries or from the opencast sites of the Ministry of Fuel and Power is usually made by the Marketing Department of the Board. This is not so in all cases because of the continued existence of a number of sales agents whose agency contracts the Board were obliged to take over on nationalization taking place. In the past many colliery concerns had bound themselves by contracts giving agents the right to dispose, in some cases, of the whole output of a colliery, and in other cases of the whole or part of the output for a particular purpose or for distribution in a particular area. In 1946 exclusive sales agencies alone controlled the annual disposal of over 30 million tons of coal, 1 million tons of coke and 600,000 tons of manufactured fuel. The continued existence of these sales agencies, whether exclusive or otherwise, would have deprived the Board of executive responsibility for the first sale of a substantial proportion of their products, and would thus have impeded the planning and reorganization of colliery sales. Accordingly arrangements have been made to bring such agencies to an end, by notice or negotiation, or by the exercise of statutory powers granted by the Coal Industry Act, 1949.

In addition to making the first sale of the coal to a merchant the Board may sell direct to a buyer. If the work involved is more than putting the coal into wagons, weighing and labelling and handing it over to the main-line railway system, the Board charge for the additional service just as any other distributor would do. In this way, the Board are distributors in competition with the private enterprise distributors.

There are three elements of the distributive trade—the wholesaler (or factor), the coal merchant, and the coal dealer. It is difficult to define the duties of each, because many firms act as factors, merchants and dealers, buying from a colliery and distributing the coal in their own lorries to the householder. The factor usually buys from the Board in bulk—either in wagon loads or shiploads—and sells in smaller quantities to the merchant or large consumer. Some of the factors do not actually handle coal at all; others own ships and have their own wharves and possess plant and equipment for handling, storing, screening and bagging coal. In normal conditions factors put to good use their knowledge of alternative sources of supply and so relieve industrial consumers and merchants of the need to maintain specialist coal buyers.

The merchant buys from factors, or direct from the Board, and usually maintains depots where coal is stocked. The coal dealer serves the domestic consumers by buying from the merchants' yards and delivering the coal in bags. It is not uncommon for the coal dealer to use his lorries or carts for other haulage work during the summer months.

It is estimated that there are more than 20,000 coal merchants and dealers in the country, many of them being very small businesses. There is room for reorganization and improvement in the retail trade, particularly in the provision of better planned depots, and modern equipment for bagging and loading the coal. Before the industry was nationalized the Coal Merchants Federation and the Co-operative Union put forward new proposals for the post-war distribution of house coal. They recommended that customers should be given complete freedom in the choice of merchants, but that distribution should be licensed so as to ensure a proper service to the public, without however precluding anyone from trading who was willing to give such service. They proposed a form of price-control that would ensure economic and efficient working and prevent price-cutting. They suggested that licensing and price regulation should be undertaken by national and area bodies representative of producers and distributors, with some State control in order to ensure observance of rules and freedom from abuse. They also recommended improvements in coal handling facilities at depots and improved working conditions for the labour employed. The implementation of these proposals has been deferred, and reorganization of coal distribution has become a matter for discussion between the National Coal Board and the distributive interests concerned.

The Board's Marketing Department operates at Headquarters, Divisional and Area levels, some of the Areas being grouped together so as not to disturb established markets and traditional sources of supply. At Headquarters, Marketing Department advises the Board on the broad lines of marketing policy, plans and co-ordinates the selling activities of the Board, and deals with such national issues as the price structure and standard conditions of sale. On transport problems the Marketing Department maintains close contact with various headquarters offices of the British Transport Commission. The Department also has the difficult task of assessing future demand for particular types and grades of coal in order that Production Department may plan output in relation to consumer demands. The Divisional Marketing Departments assist Headquarters in policy matters and carry out such policies in the Divisions. In so doing, they are given considerable freedom to plan sales organizations best suited to local requirements. The actual selling of the coal is normally done at Area level and is the responsibility of Area Marketing Managers. In addition, sales offices have been opened in the main consuming areas of the country, where representatives of several Divisions are present to keep in touch with consumers and distributors, and if necessary to investigate complaints 'on the spot'.

As well as selling coal, coke and manufactured fuel, the Marketing Department is responsible for selling the large range of coal by-products referred to on page 85. The sale of the Board's clay products, such as building bricks, refractory bricks and clay pipes, is also a function of the Marketing Department.

*Coal Supplies Organization*

Shortage of coal supplies during the war made it necessary for restrictions to be imposed on coal consumption, and restrictions will continue to be necessary in the post-war period until such time as supplies are sufficient in quality and quantity to enable normal trading to be resumed. In the meantime, therefore, neither consumers nor the Board have complete freedom in the buying and selling of coal. At the request of the Ministry of Fuel and Power the Board have accepted temporary responsibility for administering the Government's coal allocation scheme which was introduced during the war. For this purpose certain of the senior officials of the Board's Marketing Department, both at Headquarters and at Divisions, have been appointed Coal Supplies Officers. In this capacity they act on behalf of, and derive their powers from the Minister of Fuel and Power, and it is their duty to see that coal is distributed in accordance with the directions of the Government.

The 'rationing', or allocations procedure, which is adopted is briefly as follows. The Government periodically review the coal situation and decide how the available coal supplies should be shared out. Among other things, decisions are taken as to the classes of consumers which shall have priority and as to the amount by which distributed stocks should be increased or decreased during the period under review. These reviews take place every six months and cover the following half-year, i.e. the coal summer (May-October) or the coal winter (November-April). As a result, a coal 'budget' is drawn up which shows the estimated availability of coal over the period, and allocates the global tonnages to be supplied to the different categories of consumers. The National

Coal Supplies Officer (who is the Board's Director-General of Marketing) then becomes responsible for putting the Government's decisions into operation.

The tonnages of coal allocated to industry have to be divided among individual consumers. In the main this work is done by Programming Authorities situate in London, Manchester, Birmingham, Bristol and Sheffield, who work under the control of the National Coal Supplies Officer. (In the Scottish, Northern, and South Western Divisions, the programming work is done in the Board's marketing offices, and there is no separate Programming Authority.) For administrative reasons the Headquarters of Marketing Department act as Programming Authorities for gas-works, power stations, railways and coke-ovens. The allocation of house coal to depots is carried out by a different organization—the House Coal Distribution (Emergency) Scheme—which is staffed mainly by the distributive trade, and is operated by the trade on behalf of the Ministry of Fuel and Power. The principle underlying the allocation of house coal is, however, no different from that for all other classes of consumer. As a result of the allocations procedure, the weekly tonnage of house coal due for supply to each depot is determined in just the same way as is the weekly tonnage of coal allocated for supply to an individual gas-works, power station or industrial consumer. In the case of house coal, however, the Scheme is responsible for dividing the allocations to depots among the individual merchants trading from these depots. In due course, when the allocations of each consumer and the Divisional source of supply have been settled, the Programming Authorities tell each of the Divisional Coal Supplies Officers (who are the Board's Divisional Marketing Directors) what the allocations from his Division are to be. It is then the business of the Divisional Marketing Departments to arrange for the sale and supply of the coals covered by these allocations in the normal commercial way.

Coal exports are an important contribution to the national economy, and the Board are not free to sell coal abroad how and where they please. A large proportion of the total coal exports are earmarked for countries participating in the European Recovery

Programme, and the Government decides how much coal shall be shipped to each country. This decision depends on where the coal exports can give the best return in terms of food and raw materials, and where they can save dollars. Bilateral agreements are concluded between the Governments, and to implement these agreements the Board's Marketing Department then negotiates contracts in the normal commercial way either with foreign buyers, or with British exporters acting on behalf of foreign buyers.

Shortage of coal supplies in the post-war years has resulted in the establishment of two international organizations concerned with coal supplies in Europe. The main object of these organizations is to recommend the best use of available European coal resources. One of these bodies is the Coal Committee of the Economic Commission for Europe, successor to the European Coal Organization which functioned in the immediate post-war period. This Coal Committee, which is representative of all European countries in the United Nations Organization, makes a quarterly examination of the amount of coal likely to be available in Europe, and recommends an allocation of this coal among European countries. The Committee is also undertaking a long-term study of European coal production and utilization, and is seeking to formulate a uniform system of coal classification. The other body is the Coal Committee of the Organization for European Economic Co-operation. This Committee is representative only of those European countries which participate in the Marshall Plan. The Committee is concerned with long-term programmes of coal supplies and requirements, and the progress made towards the aims of European recovery. The British Government are represented on both these Coal Committees, and representatives of the National Coal Board attend meeetings of the Committees as advisers to the U.K. delegates.

*Markets and Prices*

Coal is needed for home industry and domestic use, as well as for shipment abroad. Coal exports declined considerably during the war owing to the decrease in production and the loss of access to foreign markets. Shipments continued to Eire, Spain, Portugal, and South America but supplies for these markets were limited by the increased demands of home industries engaged on war work. In 1942, the Government started opencast coal working, about a million tons being produced in that year, and by 1948 coal output from opencast workings had increased to nearly 12 million tons. However objectionable opencast coal working may be to local residents, it has yielded additional supplies which were invaluable during the years of shortage, and has enabled exports to be resumed in the post-war years. Recovery of the export markets has been slow, and although apparent progress was made in 1946, it was only possible by drawing on coal which should have been put into stock for the following winter. As a result there was a critical shortage of coal at home during the early part of 1947, which was accentuated by transport difficulties. During that year coal exports ceased, and coal was in fact imported from Poland and the United States. With this help, and the careful supervision of the stock-building programme during the year, exports were resumed at the beginning of 1948. By the end of the year 10½ million tons of coal had been exported and all the pre-war importing countries had received some British coal. Thus 1948 may be regarded as the first year of the post-war coal export drive.

Before the war the buyers of the largest tonnages of coal from this country were France, Germany, Denmark, Sweden, Eire and Italy, in that order. Because coal exports are now sold under bilateral agreements, it has been impossible to rebuild the export trade in strict relation to pre-war quantities. Owing to the changed economic conditions in the post-war world, the demand for coal from some countries is less than it was before the war. Hydro-electric production has been developed during the war years in many countries—Switzerland, Norway and Argentina, for example. In the immediate post-war years, owing to the coal shortage many countries increased their consumption of fuel oil and converted many of their plants for oil-burning. Coal exports have therefore to be competitive in the foreign markets and the quality of the coal must be satisfactory to the consumer. As exports are increased, the problem of supplying such a large proportion of the 'prime' qualities to foreign buyers will become more difficult, and to some extent it will continue to be necessary to subordinate home requirements to the over-riding necessity of exporting coal in order to achieve a balance of payments on the country's trade.

A declining market is that for bunker coal, i.e. the coal used by a ship on its voyage. The demand for bunker coal has been decreasing since 1914, owing to the increasing use of oil. In 1914, over 88 per cent of the world shipping was coal-fired, but by 1948 the proportion of the world shipping tonnage exclusively fired by coal was less than 22 per cent. Although many steamers are fitted for dual-firing, and can be converted to burn coal or oil, most modern ships are motor vessels propelled by diesel engines. The coal required for bunkers is a hard steam coal, able to withstand rough handling without producing a lot of 'fines'.

In this country, coal is the primary source of power; in 1948, the total consumption amounted to 192.3 million tons. Of this total, public utility undertakings consumed 67.7 million tons, made up of 28.7 million tons for electricity power stations, 24.5 million tons for gas works, and 14.5 million tons for the railways. The consumption of coal by public utilities has increased considerably in recent years, especially in the case of power stations. In 1948, coke-ovens consumed 22.2 million tons, and other industrial consumers accounted for 42.2 million tons, of which 8.6 million tons were used for iron and steel production. The collieries consumed 10.6 million tons, and free and concessionary coal for miners amounted to just under 5 million tons. Domestic consumption amounted to 29.2 million tons, together with 2.3 million tons of anthracite and boiler fuel. The balance of approximately 13 million tons, is accounted for by miscellaneous uses, including service requirements and shipments to Northern Ireland.

During the course of the war, price increases were permitted in order to meet increased wages and other costs, and the average pithead price per ton of coal, which had been 17*s.* to 18*s.* before the war, increased by 20*s.* during the war, and a further 6*s.* 6*d.* within the first year of nationalization. This war-time control of selling prices, and permitted increases in the form of flat-rate additions to existing prices, resulted in an anomalous price structure. The relative prices of different grades of coal were disproportionate to their qualities; low-grade coal cost three or four times as much as it did before the war, whereas the best coals no more than doubled in price. The price structure inherited by the Board was most unsatisfactory, and one of the major tasks of the Board's Marketing Department has been the development of a new rational price structure. Its aim is to relate the price of coal to its quality and size, so that a buyer will receive relative value for money regardless of which coal he chooses. To prepare this new price structure, a great deal of work has been necessary to evaluate the many types of coal on the lines indicated at the beginning of this chapter, and to study the effect of price changes on consumer needs. The varying costs of production and transport are factors which have had to be taken into account in the development of the new price structure.

Rising costs of production in industry generally, as well as shortage of supplies, have focussed attention upon efficiency and economy in the use of coal. During the war, Regional Fuel Efficiency Committees were set up, and similar Committees were established in various branches of industry. In this way, an effective consultative service developed, which emphasized the methods by which greater efficiency in the use of coal and other fuels would reduce the quantities consumed. Within the industry itself various bodies have been created for the purpose of research or guidance in the use of solid fuels. The Coal Utilization Joint Council reaches a wide public with advice and guidance in the use of fuel and appliances, and the British Coal Utilization Research Association undertakes research in the scientific design and operation of fuel installations so as to obtain maximum thermal efficiency. The Board give considerable financial and general support to these and similar organizations, and act in close collaboration with the Fuel Efficiency Branch of the Ministry of Fuel and Power.

Chapter 8

# THE FINANCES OF THE INDUSTRY

The purpose of this chapter is to describe the financial structure of the National Coal Board, and to show the form in which their accounts are presented to Parliament. To this end the accounts of the year 1948 are examined by way of illustration: the explanations given will have general, but not of course detailed, application to the accounts of subsequent years.

### *The Board's Accounts*

The Board are required by the Coal Industry Nationalization Act 1946, to direct their policy so as to ensure (consistently with the proper discharge of their other duties) that on an average of good and bad years their revenues are at least equal to their outgoings. The same Act requires them to keep proper accounts and related records, and to prepare a statement of accounts for each financial year. The Board's financial year ends on December 31. The accounts must be prepared in accordance with a direction given by the Minister of Fuel and Power, and this direction is printed in full in the Board's Annual Report for 1947. The form of the accounts must conform with the best commercial standards and distinguish colliery activities and each of the main ancillary activities of the Board. These other activities are coal-selling depots; coke-ovens; secondary by-product plants; manufactured fuel plants; briquetting plants; brickworks and tileworks; railway wagons; wagon repair shops; houses; estates and farms.

The Board's published accounts in fact go far beyond the statutory requirement of 'best commercial standards', particularly in the analysis of income and expenditure, the detail given as to fixed assets and profit and loss on the main activities carried on by the various Divisions, and the extensive supplementary information provided. The published accounts consist of a simplified

Balance Sheet and Profit and Loss Account, supported by a number of Schedules and explanatory notes showing in detail the various assets and liabilities and the transactions relevant thereto. Detailed Profit and Loss Accounts show the net amount of operating profit or loss on collieries by Divisions and Areas, and the trading results of the various ancillary activities. A full analysis is given of income and expenditure on revenue account and of general expenses incurred, irrespective of the activities to which they relate.

The accounting systems which the Board took over from the former owners were diverse, and in many cases fell considerably short of the standards normally observed in modern industrial concerns. This was particularly the case with stores control and cost accounting. The Board have not sought to impose standardized accounting systems on the whole industry but have given the Divisions and Areas discretion as to the accounting arrangements by which they produce in standard form the information needed for making up the Board's accounts as a whole. The principal accounts are maintained in the Area offices, though some accounting on wages and stores takes place in the offices of collieries or groups of collieries. Financial supervision and the aggregation of accounts is undertaken at Divisional and National levels. Mechanized accounting systems are used where it is advantageous to do so, and considerable emphasis is placed on modern methods of financial control. Such controls are essential in a large-scale organization in which the profit motive does not operate fully and those responsible for management cannot possibly have detailed knowledge of every circumstance of the various activities undertaken.

The Board's financial officers insist upon carefully prepared estimates of the financial results of proposed schemes; budgetary control to ensure that expenditure is devoted only to the purposes authorized; and close examination of the financial results of schemes undertaken in order to determine whether the financial results envisaged have been attained. Monthly profit and loss accounts for the collieries and the ancillary undertakings are scrutinized by the Area General Managers and kept under constant review at Headquarters and in the Divisions concerned.

The Board maintain 47 Area Banking Accounts at the London offices of the principal Banks, and these are used for normal business transactions. The daily balances on these accounts are cleared to a Central Banking Account at the Bank of England and this account is used for transfers to and from the Exchequer on behalf of the Minister of Fuel and Power. If on any day the Board's account at the Bank of England increases beyond a prescribed maximum the excess is transferred to the Exchequer; if the balance falls below a prescribed minimum transfers are made from the Exchequer to the account. This system of linked banking accounts with central clearing makes it unnecessary to maintain and control balances on a large number of unit accounts, eliminates overdraft charges, and keeps at a minimum the interest payable on loans from the Minister of Fuel and Power.

The Board are not exempt from payment of Income Tax but by reason of their trading loss in the first year they have not so far incurred liability for Income Tax. They are also liable for Profits Tax, and in assessing their liability to this tax they will not be allowed to charge against taxable profits the interest on their initial capital except to the extent that the capital of the former owners consisted of debentures or other loan capital.

The Board's accounts are audited by professional auditors appointed by the Minister of Fuel and Power and a copy of the Board's statement of accounts and of the auditors' report is laid by the Minister before Parliament. The Accounts are examined by the Public Accounts Committee, and in the course of this examination representatives of the Ministry of Fuel and Power and of the Board attend before the Committee for the purpose of answering questions relating to the Accounts. The Board maintain a system of internal audit, the purpose of which is not merely to prevent and detect fraud, but also to ensure compliance with financial policy and procedure and to evaluate the effectiveness of various schemes of financial control.

### *Capital, Liabilities and Assets*

A public company normally finances capital expenditure by subscriptions from shareholders or by borrowing on debentures or other securities. The Board have no fixed capital and do not issue stock, and if they wish to borrow money other than temporarily (e.g. overdrafts) they must do so from the Minister of Fuel and Power. This arrangement is advantageous to the Board in one respect: money is borrowed only as and when it is immediately needed, in the manner described on page 142, whereas an issue of shares or debentures by a public company may result in more money being taken up than is needed at a particular time.

The Nationalization Act enables the Minister of Fuel and Power to make advances to the Board, not exceeding in the aggregate £150 million up to mid-1951 and thereafter such amount as Parliament may decide. These advances are for expenditure chargeable to capital account (i.e. expenditure on development and reconstruction) and also for the provision of working capital (i.e. money necessary to finance the Board's activities between the time payments have to be made—for example, on wages—and the time when the corresponding receipts—for example, from coal sales—are to hand). The Board may, with the consent of the Minister, borrow temporarily elsewhere by way of overdraft or otherwise such sum (not exceeding in the aggregate £10 million) as they require for the purpose of meeting their obligations and discharging their functions.

The Nationalization Act requires the Board to reimburse the Minister the amount of the expenses and liabilities incurred (a) in making compensation for (i) the transfer of assets to the Board (ii) increased overhead expenses caused by severance and capital outlay refunds (see page 50), and (iii) the winding-up of coal selling schemes and a mines drainage scheme; and (b) in the issue of stock in exchange for Coal Commission Stock and in making interim income and interest payments pending the payment of compensation.

The Board are required to repay to the Minister his advances for capital development together with interest thereon. Interest is at the rate of $2\frac{1}{2}$ per cent for advances made in 1947, 3 per cent for advances in 1948, and at rates to be determined by the Treasury for advances to be made in subsequent years. Equal annual payments of principal and interest are spread over periods of 50 years and so it may be said that this generation is making a present of the coal industry to its grandchildren. In 1948 the Board financed £25 million of capital expenditure from their own resources (depreciation funds, etc.), thus saving considerable interest charges.

The amount of compensation to be paid for the ancillary undertakings and for some other items has yet to be valued by the District Valuation Boards, and so the National Coal Board do not know the exact extent of their capital liabilities. The initial capitalization of the Board is, however, estimated at approximately £400 million, and this will carry interest charges of £12 million or more per annum.

The Board's capital liabilities may be summarized as follows:

| | £ | |
|---|---|---|
| Collieries ... ... ... ... ... | 164·7 | million |
| Ancillaries (outside global sum)—Amount unknown | | |
| Minerals ... ... ... ... ... | 78·5 | ,, |
| Main-Line Railway Wagons (provisional) ... | 12·3 | ,, |
| Capital Outlay Refunds ... ... ... | 17·8 | ,, |
| Stocks of Products and Stores (provisional) ... | 39·6 | ,, |
| Advances by Minister for Capital Expenditure ... | 41·1 | ,, |
| Total (so far as at present ascertained) ... | 354·0 | ,, |

By December 31, 1948, capital repayments had reduced this amount to £343.8 million.

The Board's deferred liabilities at the end of 1948 amounted to £31.6 million, and of this £28.3 million was for workmen's compensation payments. The Nationalization Act did not deal with the transfer of liability for workmen's compensation payments to workmen employed in activities which were taken over by the Board, but left the matter to be settled by negotiation between the Board and the former owners; an agreed amount plus interest will be paid to the Board. The Board's accounts contain the provisional item of £16.7 million for the liability taken over from the colliery concerns and others (less compensation payments made since nationalization) and also for the probability of the Board making payments greater in some cases than those which the former owners might have been expected to make. £11.5 million is provided for cases arising between the primary vesting date and the date on which industrial injuries insurance began to operate. Other deferred liabilities for which provision is made are the Board's supplementary injuries scheme; surface damage resulting from mining operations; compensation for loss of office by staff formerly employed in the coal industry; and the rebuilding or relining of coke-ovens and kilns. As the Board have decided to carry their own insurance risks (with certain exceptions) an Insurance Fund is being built up to meet claims arising, and this fund constitutes part of the deferred liabilities.

The nature and extent of the Board's current liabilities are shown by the 1948 Balance Sheet. Of a total of £49.6 million the sum of £16.3 million represents what is owed by the Board for supplies and services provided towards the end of the financial year. £13.5 million is the amount due to the Minister of Fuel and Power for the year's payments to former owners in lieu of interest on outstanding compensation. £8.8 million represents charges, mostly wages, which accrued due at the end of the financial year but will not be paid until the new financial year; £2.1 million is provided for the Board's liability to their Superannuation Fund. £4.1 million is the amount due to creditors not otherwise accounted for, and includes what is estimated to be owed to the Minister of Fuel and Power for colliery plant which the Board have taken over from him, but for which a settlement has not yet been reached. The balance of current liabilities is made up of £3.1 million due to the Inland Revenue authorities for tax deducted from wages and salaries and from rent and interest payments; bank overdrafts, and a few unsettled apportionments surviving from the vesting date.

The remaining item on the liabilities side of the 1948 Balance Sheet is the Reserve Fund, amounting to £3.1 million, the nucleus of which was taken over from the Coal Commission. The Board are required by the Nationalization Act to establish a Reserve Fund which they may apply as they think fit, though not otherwise than for the purposes of the Board. The Minister may give directions to the Board (with Treasury approval) as to any matter relating to the establishment or management of the Reserve Fund, and as to carrying sums to the credit of the Fund or applying the Fund. No allocations were made to the Reserve Fund during 1947 and 1948 except for a small profit made in 1948 on the realization of investments of the Fund.

The principal fixed assets of the Board are the collieries and associated properties which were transferred to them, their value being the 'global' sum of £164.7 million. This value is augmented by the approved capital expenditure, amounting to £17.2 million incurred by the Colliery concerns shortly prior to nationalization. (The difference between this and the corresponding item of £17.8 million shown under liabilities is accounted for by Capital Outlay Refund items in stock at the date of nationalization.)

The value of most other fixed assets, in particular the ancillary assets such as coke-ovens, manufactured fuel plant, brickworks, houses, farms, etc., will not be known until they are determined by the District Valuation Boards. The value of main-line railway wagons taken over from the collieries has, however, been provisionally determined at £12.3 million, as these were transferred to the British Transport Commission on nationalization of the railways. The other principal assets are the minerals taken over from the Coal Commission, and their value (including some other items of Commission property) is shown in the 1948 Balance Sheet at £78.5 million. The Board's fixed assets also include such additions as they have themselves made by purchase and development, and these are represented by £43.9 million in the 1948 accounts. They consist for the most part of plant and machinery, mines (including surface works), and houses and office buildings. Disposals amount to £13.6 million, this figure including more than £12 million for the disposal of the railway wagons.

Provision for depreciation is complicated by the fact that the total value of the fixed assets is not yet known. The cumulative provision for depreciation at the end of 1948 amounted to £32.1 million.

The Balance Sheet contains a suspense item of £1.9 million which represents the proportion of holiday payments to workmen in 1947 which accrued before January 1 of that year and so is claimed by the Board to be a liability of the former owners. The liability is disputed by the Mining Association.

Among the Board's assets are investments of £0.2 million, the greater part of which consists of debentures of the British Coal Utilization Research Association, whose operations the Board finance to a substantial extent. These debentures represent advances to the Association for capital expenditure on development schemes. Current assets consist of stocks of products, stores and materials which were valued at the end of 1948 at £59.9 million; trade and other debtors and bills receivable amounting to £71.3 million, and £2.1 million in cash and bank balances. Provision for debts due to the Board includes a sum of £15.8 million due from the former owners for workmen's compensation payments.

A simplified version of the Board's 1948 Balance Sheet is as follows:

| *Liabilities* | £m. | *Assets* | £m. |
|---|---|---|---|
| Capital Liabilities (so far as ascertained) ... ... | 343·8 | Fixed Assets (so far as ascertained) ... ... | 270·8 |
| Provision for Deferred Liabilities ... ... | 31·6 | Suspense Account (1947 Holiday Pay) ... ... | 1·9 |
| Current Liabilities ... | 49·6 | Investments ... ... | 0·2 |
| Reserve Fund ... ... | 3·1 | Current Assets ... ... | 133·4 |
| | | Deficit ... ... ... | 21·8 |
| | 428·1 | | 428·1 |

The deficit shows the amount by which the Board's liabilities exceeded their assets. It may be regarded as a debt which the Board's revenue accounts owe to the capital account. In a more general sense the deficit is a liability which the Board will have to dispose of by future trading. In the meantime the loss has to be made good from the Board's own resources and this is done by making temporary use of the provisions for deferred liabilities.

### *Trading Results*

The deficit of £21.8 million consists of expenditure amounting to £0.2 million incurred by the Board in 1946 before trading commenced; a loss of £23.3 million incurred during the first year's working; and a surplus of £1.7 million which resulted from activities during the second year of nationalization. The loss of £23.3 million in 1947 was made up as follows;

| | £ | |
|---|---|---|
| Operating Loss ... ... ... | 6·2 | million |
| Loss on Imported Coal ... ... | 1·7 | ,, |
| Compensation for Loss of Office ... | 0·4 | ,, |
| Interest and Interim Income ... ... | 15·1 | ,, |
| | 23·4 | ,, |
| Miscellaneous Income ... ... ... | 0·1 | ,, |
| | 23·3 | ,, |

The operating loss of £6.2 million resulted from a loss of £9.2 million on collieries, less a profit of £3.0 million on ancillary activities. Not all the Divisions made losses in operating the collieries, and the results varied from a profit of £7.5 million in the East Midlands Division to a loss of £10.7 million in the South Western Division. On balance a total profit from colliery working of £22.0 million had to be set against a loss of £31.2 million. 'Losing' collieries were no new thing in the industry, and various devices, including the Coal Charges Account (see page 33) had been used in the past to make the profitable areas pay for the losing areas. If the Board had been able to close unprofitable collieries the financial result of the year's working would have been different; but for many reasons, in particular the imperative demand for coal, such closures could not be made on any significant scale.

The changes which the Board were expected to introduce, and did in fact introduce, in the first year of nationalization inevitably required considerable expenditure to be undertaken in advance of the increased productivity towards which it was directed. Capital expenditure (which amounted to £19 million in 1947 and £25 million in 1948) was mainly for additions to capital assets, or their replacement or improvement, and could not be expected to show an immediate return. Concessions to workmen in order to secure their co-operation and good will, as well as to attract recruits, did have results apparent in the first years of nationalization in the form of increased output of coal, reduced absenteeism, better output per manshift, and an increase in the number of workers employed in the industry. These concessions, in the form of the five-day week, increased minimum wages, and other benefits, as well as the increased cost of materials, stores, repairs and power, resulted in the colliery cost per ton of coal being 4*s.* 3*d.* greater in 1947 than in 1946. This was an increase of 12.8 per cent and was in fact no greater than the general increase of costs in British industry in the year 1947. The need to increase the price of coal was apparent, but, under an informal undertaking they had given, the Board were not free agents; they were obliged to obtain the approval of the Minister of Fuel and Power to whatever increases they thought were necessary, and in fact the price increases which the Board proposed were scaled down by the Government and were not introduced as early as the Board had intended.

Notwithstanding further increased costs of production in 1948 the collieries in that year showed on balance a profit of £16.2 million. The profit of the East Midlands Division increased to £14.5 million and the loss of the South Western Division fell to £5.7 million; the North Western Division turned a 1947 loss of £1.6 million into a small profit of £0.03 million. As a result of inland price increases decided upon in 1947 colliery proceeds for 1948 improved by more than 5*s.* a ton, and they had the benefit of a further 1*s.* 6*d.* a ton attributable to the fact that coal exported sold at a higher price than in the home market.

In 1947 the Board's ancillary activities made a profit of £3.0 million, principally on coke-ovens and railway wagons. In 1948 this profit fell to £1.2 million, the difference being accounted for by (a) the transfer of railway wagons to the British Transport Commission (b) smaller profit on coke-ovens in consequence of increased provision for rebuilding and depreciation, and (c) a heavy loss (£0.7 million) caused by the need to make good considerable arrears of repairs to houses.

The 1947 loss of £1.7 million on coal imported from America and Poland was the unavoidable consequence of a Government decision enforced by the necessity of increasing the supplies of coal available so as to prevent another winter fuel crisis.

The following is a summary of the trading results for the first two years of nationalization before charging interest and interim income payments, losses being bracketed thus [ ].

| | 1947 | 1948 |
|---|---|---|
| | £ | £ |
| Collieries ... ... ... ... | [9,203,905] | 16,172,694 |
| Coal Selling Depots ... ... | 58,515 | 103,191 |
| Coke-Ovens ... ... ... | 925,555 | 536,705 |
| Secondary By-Product Plants ... | 147,207 | 174,752 |
| Manufactured Fuel Plants ... ... | 57,996 | 57,405 |
| Briquetting Plants ... ... ... | 228,409 | 236,640 |
| Brickworks and Tileworks ... | 122,847 | 191,073 |
| Railway Wagons ... ... ... | 1,203,407 | 149,677 |
| Wagon Repair Shops ... ... | 111,491 | 56,284 |
| Houses ... ... ... ... | [202,653] | [662,840] |
| Estates and Farms ... ... ... | 90,545 | 76,996 |
| Other Activities ... ... ... | 218,604 | 299,182 |
| Licensed Mines ... ... ... | 54,682 | 123,197 |
| | [6,187,300] | [17,514,956] |

Although the first two years working show a net loss of £21.8 million, the Board made payments to the Government during those two years of £30.7 million in respect of interest and interim income payments on compensation. These payments have to be made whether or not the Board make a profit, and this obligation is a further example of the difference between the Board and a joint-stock company, for a company would not distribute any dividend in a bad year. Such a company would, of course, have to pay interest on debentures, etc., but any normal company would have a substantial proportion of its capital in ordinary shares, and on these the dividend could fluctuate.

### *The Cost of Coal*

In terms of money the average cost of producing a ton of coal increased from 16*s.* in 1938 to 37*s.* in 1946, 41*s.* 3*d.* in 1947 and 45*s.* 7*d.* in 1948. These increases are almost entirely accounted for by wages and allowances, which increased by about 21*s.* 6*d.* a ton during the period 1938-1948, and by industrial materials and power used in coal production, the cost of which in 1948 was nearly 2½ times as great as in 1938. These increased costs were not offset by increased production, and in fact the production of deep-mined coal decreased from more than 230 million tons in 1939 to less than 175 million tons in 1945. From 1946 onwards output has slowly but steadily increased, and by the end of 1948 the rising costs of production had begun to abate, and have since started to fall.

The wages element in the 1947 increased cost of production did not have effect over the whole year, and so cost only £24 million in that year, compared with £62.5 million in a full year. The 1947 increased cost of 4*s.* 3*d.* per ton as compared with 1946 was made up as follows:

| | *s.* | *d.* |
|---|---|---|
| Increased Wages and other benefits to Workmen ... | 2 | 6·4 |
| Increased Cost of Materials, Stores, Repairs and Power | | 11·9 |
| Superannuation and other benefits to Officials and Staff | | 4·2 |
| National Insurance (Statutory Increase) ... ... | | 1·5 |
| Recruitment, Training, Hostels and Scientific Expenditure | | 1·5 |
| Net balance of other Increases and Decreases ... ... | | 1·6 |
| | 4 | 3·1 |

In 1948 the element of wages and allowances in production costs again increased by about 2*s.* 9*d.* a ton, but this was mainly in consequence of the changes introduced in 1947. The cost of materials and power again increased by 1*s.* a ton, and additional costs were also incurred for national insurance and the supplementary injuries scheme.

With figures averaged over the country as a whole, the detailed costs of producing a ton of coal in 1948 were as follow:

| | *s.* | *d.* |
|---|---|---|
| Wages (including allowances in kind) ... ... ... | 28 | 3·8 |
| Holiday Pay ... ... ... ... ... ... | | 11·2 |
| Supplementary Injuries Scheme ... ... ... | | 2·6 |
| Workmen's Compensation ... ... ... ... | | 5·0 |
| National Insurance ... ... ... ... ... | | 7·5 |

| | s. | d. |
|---|---|---|
| Roof Supports ... ... ... ... ... ... | 2 | 9·6 |
| General Stores ... ... ... ... ... | 3 | 0·1 |
| Repairs and Renewals ... ... ... ... ... | | 11·0 |
| Power, Heat and Light ... ... ... ... | 3 | 2·5 |
| Salaries ... ... ... ... ... ... ... | | 11·7 |
| General Expenses ... ... ... ... ... | 1 | 9·2 |
| Administrative Expenses ... ... ... ... | | 10·4 |
| Depreciation ... ... ... ... ... ... | 1 | 6·0 |
| | 45 | 6·6 |

It will be seen that wages, holiday pay, the supplementary injuries scheme, workmen's compensation and national insurance account for 67 per cent of the cost. Roof supports, general stores, repairs and renewals, power, heat and light, and depreciation, account for 25.1 per cent. Salaries and administrative and general expenses amount to 7.9 per cent. The general expenses are made up of such items as superannuation, rent and rates, welfare levies and grants, grants to pithead baths, wagon charges, printing and stationery, surface damage claims, insurance provision, etc., etc.

The pithead proceeds per ton of coal saleable, amounted in 1948 to 47*s.* 2.5*d.*, yielding a profit of 1*s.* 7.9*d.* per ton. After taking into account the other activities of the Board, and paying interest and interim income, a profit of £1.7 million was available for the purpose of reducing the deficit of £23.3 million brought forward from 1947.

## SUGGESTIONS FOR FURTHER READING

A useful picture of the British coal industry just prior to nationalization is contained in *The British Fuel and Power Industries*, a report by P.E.P. (Political and Economic Planning) published in 1947. More detailed background material is available in *Report on the British Coal Industry* by P.E.P. (1936), and *Report of the Royal Commission on the Coal Industry* (the Samuel Commission) published by H.M. Stationery Office in 1925.

The *Report of the Advisory Committee on Coal Mining* (The Reid Report) published by H.M. Stationery Office in 1945 is full of valuable information on the technicalities of coal mining. The Report has a useful glossary of mining terms.

The organization and development of the nationalized industry by the National Coal Board are fully described in the Board's *Annual Reports*, commencing with the year 1946, and published by H.M. Stationery Office. These reports are eminently 'readable'.

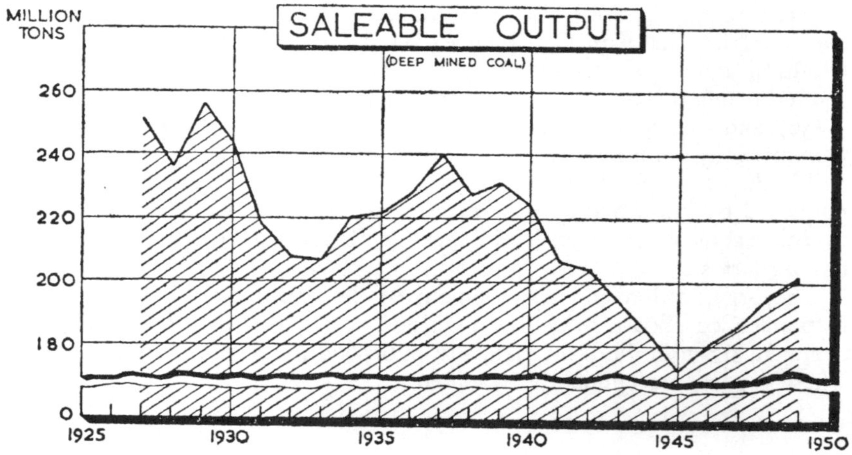
MILLION TONS
SALEABLE OUTPUT
(DEEP MINED COAL)
260
240
220
200
180
0
1925
1930
1935
1940
1945
1950

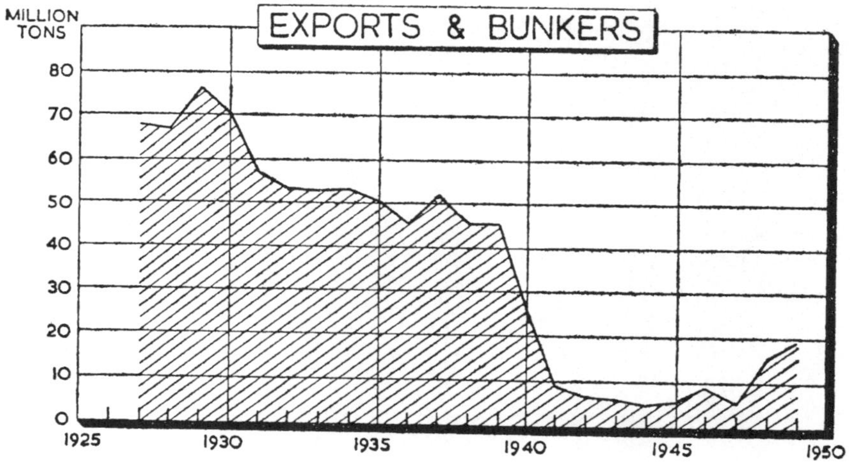
MILLION TONS
EXPORTS & BUNKERS
80
70
60
50
40
30
20
10
0
1925
1930
1935
1940
1945
1950

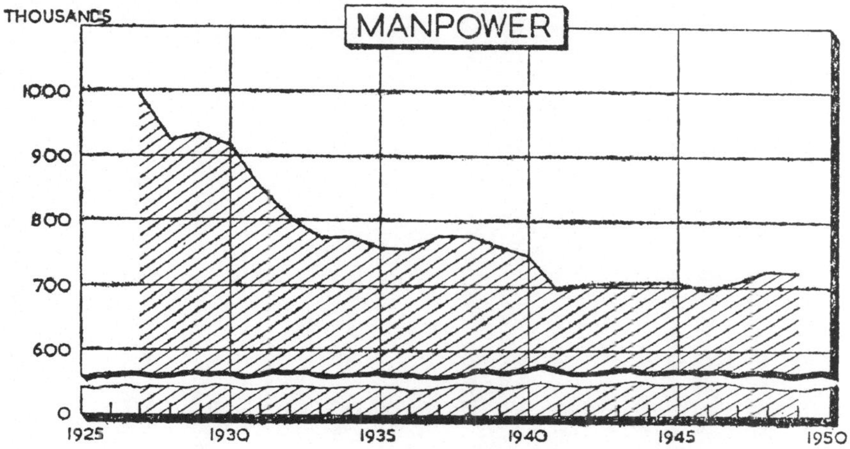


This graph shows wage earners on colliery books, excluding clerks and salaried persons.

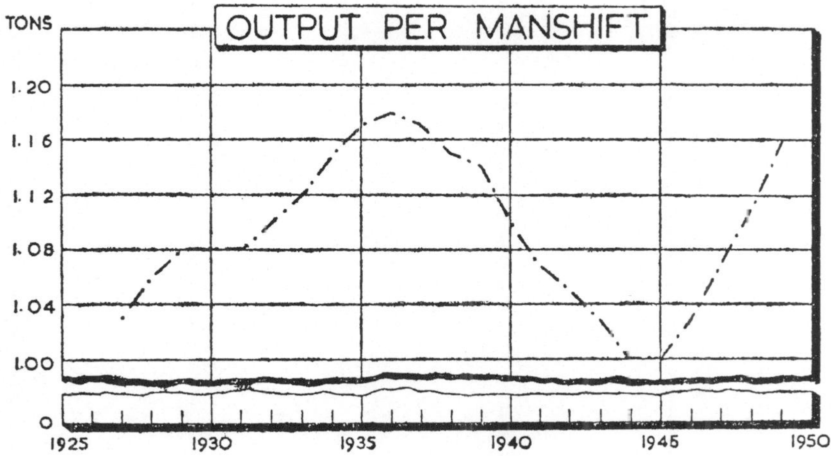

# INDEX